THE CHANNEL TUNNEL STORY

THE
CHANNEL
TUNNEL
STORY

Michael R. Bonavia
MA, PhD, FCIT

DAVID & CHARLES
Newton Abbot   London   North Pomfret (Vt)

For Susan

British Library Cataloguing in Publication Data

Bonavia, Michael R.
  The Channel tunnel story.
  1. Tunnels—English Channel
  I. Title
  385'.312   HE380.G7

  ISBN 0-7153-8964-5

Photoset in Linotron Optima by
Northern Phototypesetting Co, Bolton
Printed in Great Britain
for David & Charles Publishers plc
Brunel House Newton Abbot Devon

Published in the United States of America
by David & Charles Inc
North Pomfret Vermont 05053 USA

Contents

List of Illustrations

Cessons d'écrire des articles sur le
'Channel Tunnel', creusons le sol avec
des machines puissantes – AGISSONS.

*(Let us stop writing articles about the
Channel Tunnel and start digging with
powerful machines – ACTION!)*
(Le Monde Industriel et Commercial, 1917)

Introduction and Acknowledgements

The Channel Tunnel has been described as less an engineering project than a state of mind. Certainly most British people who think about it seem to have been born with a strong bias, either in favour or against, and nothing they hear or read later is likely to change this. At times, opposition seems not merely irrational but emotional – the almost tearful pleas against the tunnel of Sir Garnet Wolseley and Admiral Sir William Horsey in the last century were extraordinary, coming from otherwise hard-bitten serving officers. In France the attitude has always been far more practical. Memory recalls a visit to the Calais area in the 1970s, at a time when some residents of Kent were being fanatical in their opposition to the tunnel. A French farmer was met, walking his fields on the actual site of the tunnel entrance. Asked how he felt about having his land expropriated he shrugged his shoulders in the inimitable French way and replied 'Why should I care? I shall be well compensated and may buy another property in the neighbourhood. And the tunnel will be a good thing for both France and Britain'.

It cannot be said that a pro- or anti-tunnel attitude follows political party lines. In 1964 a Conservative Minister (Ernest Marples) announced a decision in principle to go ahead. A couple of years later a Labour Prime Minister, Harold Wilson, repeated the commitment. In 1970 a Conservative administration approved the tunnel; in 1975 a Labour administration suddenly abandoned it. Why has it been such a political football?

The great Alpine tunnels – the Mont Cenis, the St Gotthard, the Simplon, the Arlberg – were all completed under conditions of appalling difficulty. Most were excavated through rock, involving the continuous use of explosives. Most also encountered serious inflows of water, and sometimes temperatures rose so high that work became almost impossible. Many lives were lost in the work. Yet today it is unthinkable that France, Italy, Switzerland and Austria could do without these links, especially since the railway tunnels have been joined by the great highway tunnels such as the Mont Blanc. Before the rail tunnels were built, travellers had to make long journeys by horse-drawn coach over the mountain passes – when the weather permitted. In the case of the Mont Cenis, whereas the railway tunnel is 8½ miles long, the road journey was then almost 50 miles in length. By comparison, the Channel Tunnel offers far fewer construction dangers and problems than the Alpine tunnels, considered as an engineering task; it passes through an almost ideal tunnelling stratum. Why was it not built years ago?

This book seeks to describe some of the curious vicissitudes that the tunnel has undergone in over a century of serious promotional effort, with particular emphasis on the railway component. Writing it has recalled to mind many experiences during the seven years, ending in 1974, during which I headed the British Rail planning team on the Channel Tunnel. It has also been fascinating to retrace the early history of the tunnel. Moreover, at the date of writing, the omens seem good (apart from some political cataclysm) for actual construction to start soon. But there have been, and continue to be, many books written on the tunnel; consequently I have deliberately concentrated upon the railway aspects of the project, with the involvement first of the South Eastern Railway, then that of the Southern Railway, and lastly that of British Railways.

Various friends and former colleagues have generously helped with the loan of documents and with personal recollections. I am particularly grateful to Mr Anthony Bull, CBE, for access to the papers of the late Sir William Bull, now in the Archive Centre of Churchill College, Cambridge, and for permission to quote from them; also to Mrs Lorna Corin, for information and the loan of papers about her grandfather, Sir Arthur Fell. Mr J. L. Harrington has been extremely helpful regarding the Southern Railway's attitude, and in particular that of Sir Herbert Walker. He has also allowed me to see some Channel Tunnel archives in his possession.

Lady Valentine has very kindly given me access to the papers of her late husband, Sir Alec Valentine, a former member of the Channel Tunnel Study Group. Mr C. A. Tysall, a former associate in the BR planning team, has been unsparing in helping with technical and engineering questions, as well as lending me various key documents. Another former associate, Mr D. E. Mitchell, has given valuable help. Two friends in Channel Tunnel matters, Lt Commander Christopher Powell, RN, and Mr Donald Hunt, have also been informative and helpful. I am moreover indebted to Messrs Weidenfeld for permission to quote some extracts from *The Castle Diaries* in Chapter 14.

On the current position, my latest successor as Director, Channel Tunnel, at BR headquarters, Mr Malcolm Southgate, has been of much willing assistance. Throughout the developments in the 1975–1985 decade, I have been greatly indebted to Mr D. P. Williams for a flow of information. An important paper presented by Mr R. G. Pope, formerly a member of the BR Channel Tunnel Department, to the Institution of Railway Signal Engineers has provided much of the information set out in Chapter 11. My thanks are due to Mr Pope and the Institution. I am also greatly indebted to Mr A. F. Gueterbock of Eurotunnel for checking Chapter 16.

The literature of the Channel Tunnel is vast; two substantial bibliographies of the Channel Tunnel are to be found in the Guildhall Library, London. It would be pointless to attempt to reproduce them here, even on a highly selective basis. Primary sources are to be found in the Public Record Office at Kew and in the archives of the British Transport Commission now in the custody of the British Railways Board. Later files of the British Railways Board covering more recent years will no doubt be available to historians in due course. Other sources are papers of Channel Tunnel Investments Ltd, and there is a wide field of study open to any bi-lingual researcher who may wish to review the tunnel's history from the French side. Transport historians who select some particular aspect or episode in the tunnel's long history will be able to make prolonged explorations among these primary sources and produce monographs peppered with source references, which would be inappropriate in a general history such as this, covering more than a century and a half.

Although friends and former colleagues have read various chapters and helped to eliminate errors, the responsibility for any continuing mistakes is mine alone.

Michael R. Bonavia
Haslemere, 1987

1 The Challenge of the Channel

Millenia ago, England was joined to the Continental land mass along what is now the South Coast, though it seems that a major river, often now called the Old Rhine, then flowed westward into the Atlantic. Eventually the rise in the levels of the oceans following the ending of the Ice Age and the melting of the Polar ice caps allowed the waters of the Atlantic and the North Sea to submerge much of the land, until only an isthmus, sometimes called the Wealden Island, joined England to Europe. It was over this isthmus that much of the flora and fauna of Europe is supposed to have colonised Britain.

Finally, the process of separation was completed some 7,000 to 8,000 years ago when the North Sea and the Atlantic finally broke down the isthmus and mingled their waters. The chalk formations of the North Downs and of the Pas de Calais slipped, to expose the white cliffs of Dover and the well-known headlands in France, Cap Gris Nez and Cap Blanc Nez. The chalk contains three main strata, named in English, Upper, Middle and Lower; in French, Sénonien, Turonien, and Cénomanien. In France the Lower Chalk is also known as Craie Grise or Craie de Rouen.

The depth of water in the Strait reaches maxima of 180 to 220ft. The three layers of chalk below the sea bed have a total thickness of about 260ft on the English side, about 210ft on the French side. Below the chalk is gault. The Lower Chalk is easily cut by machinery but it can stand without support and is largely impervious to water.

The Channel experiences some unique tidal streams, which can combine with a strong wind to produce extremely difficult navigational conditions. Heavy sea-fogs are also apt to arise with disconcerting suddenness.

There are two humps in the sea bed known as the Varne and Colbart banks. The Varne is a submerged ridge, about 30ft below the surface at high tide and much less at low tide. It is midway between Cap Gris Nez and Dover and is marked by a lightship; the Colbart is slightly further to the South.

Just twenty-one nautical miles separate these shores; and for centuries men have stared at the opposite coast, so tantalisingly near and yet so hard to reach in safety on those all too frequent occasions of stormy weather, when wind and tide co-operate to whip up the waves – or when the Channel fogs blot out everything from view. Between November and February, winds of Force 7 or 8 can be expected on between nine and fifteen days in a month.

England's white cliffs are more conspicuous from France than is the French shore from Britain. In fine weather Dover's residents, however, can enjoy reading the time of day through a telescope focussed on the town hall clock in Calais.

Across this narrow strip of water both traders and immigrants have moved throughout the centuries. But the Dover Strait is not just a water barrier between England and France; it is a vital corridor linking the Atlantic with the North Sea and the Baltic. The overseas trade of Russia, Finland, Norway, Sweden, Denmark, Germany, Holland, Belgium and France passes along it, at right angles to the Anglo-French traffic stream. It is often described as the busiest waterway in the world, with hundreds of ship movements every day.

On the British side, the Strait has no natural harbours with safe anchorages such as Chichester Harbour or Southampton Water. At Dover, the small river Dour flowing down from the chalk hills brought down, over the years, silt and shingle which formed a natural breakwater of sorts. At Folkestone the even smaller stream coming down through the Foord gap also built up a spit of shingle that gave some shelter for small vessels. Rather similar conditions applied at Calais and Boulogne on the French side. Over the centuries, these natural breakwaters were artificially extended and wharves built; eventually piers followed and dredging started to keep the shingle and silt from blocking the harbour entrances. British governments insisted that there must be proper communication with the Continent even before 1686, when a contract was made for carrying the mails between Dover and Calais, and Dover and Ostend, the contractor receiving £1,170 a year for the work. Wars with France frequently interrupted the service but by 1784 there were four English and four French ships providing a twice-weekly service. During hostilities the Continental mails were transferred to the Harwich–Ostend route.

After the battle of Waterloo, services via Dover were reopened, and steam propulsion arrived, anticipating its use on railways. In 1820 the *Rob Roy* paddle steamer of 90 tons started on the Dover–Calais service as a private venture, alongside the Post Office's own ships – three mail packets on the Dover–Calais route and two for Dover–Ostend. The Calais route was also served by French ships; on the Calais service each nation carried only its outward mail, but on the Ostend route the

British packets carried mail in both directions, with a subsidy from the Belgian Government.

The South Eastern Railway, which had opened from London to Folkestone in 1844, quickly purchased the harbour at Folkestone and arranged with the New Commerical Steam Packet Company to run a day service to Boulogne. Arriving in Dover in 1845, the SER formed a subsidiary, the South Eastern and Continental Steam Packet Company, which ran steamers from Folkestone and Dover to Calais, Boulogne and Ostend.

The SER gave passengers via Folkestone preferential treatment compared with those using Dover, where the transfer from train to ship was inconvenient and involved a long walk. Dover, though remaining the government's preferred port, resented the rise of Folkestone and welcomed the arrival of the London, Chatham & Dover Railway in 1861. In that year the Dover Harbour Board was created, taking over from the former Harbour Commissioners, and it embarked upon various improvements. The Admiralty pier, which had been built in 1851, was linked to both railway systems from 1864, and facilities for Continental ferry services were much improved, even though Folkestone continued to be widely used by the SER, partly because Boulogne was served better than Calais by the Nord Railway of France. For many years, even though the Nord shared the responsibility for Calais harbour with the French Government, the rail connection with Paris was by a roundabout route via Lille, until at last in 1867 the direct line over the chalk hills between Calais and Boulogne was built. The shorter rail journey in France thus helped to compensate for the rather longer sea crossing and for the delays at Folkestone caused by the need for trains to reverse and be worked up and down the very steep gradient of the Harbour branch. Until 1867 in fact Dover–Calais passengers bound for Paris often had found it quicker to take the horse-drawn diligence to Boulogne and join the train there.

The London, Chatham & Dover Railway, coming later on the scene, originally had no statutory powers to operate ships. In 1854 the government had decided to put the mail services out to contract, and the firm of Jenkins & Churchward was successful in tendering for the Dover–Calais and Dover–Ostend routes. The firm took over four former Admiralty packets and acquired several new ships, one of which, *Prince Frederick William* of 330 tons, once crossed from Dover to Calais in 83 minutes, quicker than Sealink's crossing time of 90 minutes 120 years later.

In 1862 the Churchward contract was not renewed and the service was offered by the Post Office to the South Eastern Railway, which declined it. The LCDR thereupon bid for and obtained the mail contract and arranged for Jenkins & Churchward to continue the service, until the railway purchased the firm for £120,000.

The LCDR obtained its shipping powers in 1864 and built up its own fleet, including the former Jenkins & Churchward ships. Cross-Channel traffic then started to benefit from the rivalry between the two railway companies and between the Dover–Calais and Folkestone–Boulogne routes. Dover began at last to catch up Folkestone's early development (in 1856 120,000 passengers had passed through that port compared with only 70,000 between Dover and Calais).

The ships that provided the Channel crossings grew steadily in size throughout the nineteenth century. The LCDR fleet – apart from the vessels acquired from Jenkins & Churchward, mostly of between 171 and 358 tons – began with the ominously named *Foam, Breeze* and *Wave* of 504, 349 and 393 tons respectively.

The SER had obtained its statutory shipping powers earlier, in 1853, and bought eight ships from the South Eastern and Continental Steamship Company. These were joined by the *Victoria* and the *Albert Edward* of 374 tons, in 1862; some eleven others followed, up to the fusion of the two railways in 1899, the last being the *Mabel Grace* of 1,215 tons.

The Nord Railway Company of France also operated shipping services though on a smaller scale than the British railways.

Following the transfer of the mail contract to the LCDR, the SER decided to concentrate its services for a time on Folkestone, even though its boat trains continued to serve Dover and the LCDR ships. Eventually a Continental traffic agreement between the two railways was arranged in 1863 providing a common fund of receipts, to be divided 68% to the SER and 32% to the LCDR, changing progressively to 50/50 by 1872. In 1869 it was also agreed to negotiate for the mail contract jointly, an agreement broken by the SER in 1872 when that railway grabbed the contract for itself.

The long period of inter-railway hostility finally ended in 1899 with the working union of the two systems although the independent companies, with their shareholders and boards of directors, remained in being while a statutory Managing Committee operated the combined undertaking. This long overdue reform led to development of the shipping services, with the replacement of paddle steamers by more modern screw steamers with steam turbines instead of reciprocating engines. Some people may still remember the *Riviera* and the *Engadine,* predecessors of the fleet built for the Southern Railway after 1923, in which the *Canterbury* was perhaps the best-known and best-loved ship.

But even these fine, comparatively modern vessels could not banish the dreaded sea-sickness which the short, steep waves of the Channel crossing so often produce. Victorian writers often reflected the fear of sea trips; and well they might, if one considers what cockleshells the early packets were, and the ferocity of the Channel gales. The South Eastern Railway had opened the Lord Warden Hotel at the entrance to the Admiralty Pier in Dover in 1851, where nervous travellers could, if the weather was severe, stay until the sea quietened down or, after a rough passage from France, be able to recover before resuming their journey.

Much thought and ingenuity went into attempts to reduce sea-sickness, sometimes by designs that recall the comic drawings of Heath Robinson. The *Castalia* of 1874 was built for one of the independent shipping companies which struggled to compete with the railway services – ultimately without success. This vessel was a sort of catamaran, with paddle-wheels between the twin hulls. Her speed was low, only eleven knots, and she failed to offer more stability when compared with conventional designs. She was sold and became a floating hospital isolation ward.

Soon after came the *Bessemer,* planned by the great ironmaster Sir Henry Bessemer himself. This odd vessel contained a saloon hung on a horizontal axis which could be revolved by hydraulic cylinders on each side – controlled, if you please, by a man watching a spirit-level! To be fair, the intention was to replace the man by a gyroscope driven by a steam turbine *if* the tilting mechanism worked properly. The machinery never was tested, because even with the saloon clamped in place the ship was virtually unsteerable, even in calm water. She crashed into the pier at Calais. Rather unwisely, the saloon had been lavishly and expensively decorated before the trials took place. It was taken off when the hull was sold, and adapted for life ashore as a summer-house.

Another freak was the *Ernest Bazin*, built in France, in which the hull was 'supported' on six huge rollers. The *Castalia* was followed by another and rather more effective double-hulled ship, the *Calais–Douvres* of 1878. She was also too slow for the service and was difficult to handle; she was eventually withdrawn by the LCDR in 1887. Since then the naval architects for Channel ferries have stuck to more conventional designs, the chief innovation being the Denny-Brown stabiliser which reduces rolling.

The exploits of the cross-Channel ferries in the two world wars are a major epic. Between the wars the most significant development was the introduction of the train ferry ships in the Channel (and also across the North Sea), carrying both railway wagons and the London–Paris sleeping cars of the Night Ferry. The service was developed from a ferry designed to carry war materials to France in 1918, based at Richborough in Kent. A second innovation – a foretaste of things to come – was the inauguration of a car-ferry service by the Southern Railway with the *Autocarrier* and by Townsend Bros with the *Forde*. Loading and unloading of cars was performed by cranes, rather slowly and unsatisfactorily.

The aftermath of the second world war was a turning point in cross-Channel traffic. The growth in incomes led far more people to travel abroad and the motor car explosion of the 1960s meant that road vehicle traffic became more and more important for the ships, which were thenceforth designed primarily as roll-on, roll-off (RoRo) ferries.

The Dover Harbour Board cashed in on this surge of traffic and developed the Eastern Docks specifically for the roll-on, roll-off business, without rail connections, which were now confined to the old Marine Station, renamed Western Docks. Folkestone had car ferry berths installed as well, but on a much less extensive scale than Dover.

Another striking feature was the ending of the former railway monopoly; Townsend Bros, as mentioned above, had moved into the car-ferry business in a small way between the wars, but after the second world war its services rapidly built up and soon challenged the consortium of ferry fleets calling itself Sealink, incorporating the ships of the nationalised British Railways, the French Railways and Belgian Marine (as well as the Zeeland Company of Holland). Sealink UK progressively distanced itself from the railway connection until it was privatised in 1984. Other operators such as the Olau Line, the Sally Line, and P&O entered the ferry market as well.

To-day's cross-Channel services bear little resemblance to those of the last century, even though the sea can be as rough, the wind as strong, and the fogs as dense as ever. Ferries shuttle backwards and forwards like buses in Piccadilly, their entrance to harbour controlled by green and red signals like those of a railway station. The ships are much larger than before, their bellies loaded with cars, lorries and coaches, often on two decks. Mass catering of self-service pattern has replaced the former solicitude of dining-room stewards. But – and this is astonishing – the 1986 standard Dover–Calais time of 75 minutes for Townsend Thoresen and 90 minutes for Sealink compares badly with the 72 minute run by the *Invicta* of 1882 – over a century ago! Other specimen crossing times by LCDR ships in 1897 were 77 minutes, 79 minutes, 74 minutes and – best of all – 70 minutes exactly. For many years Victoria station in London displayed an advertising slogan 'Sea Passage One Hour' – no doubt meaning pier-head to pier-head, not quayside to quayside.

Of course there are subsidiary forms of sea crossing. The Dover Strait is crossed by Hovercraft services and the Jetfoil links Dover with Ostend. But these are relatively small craft of limited traffic capacity and – most important – above a certain wave-height they cannot operate safely.

Despite all the modern advances in technology, the sea can still sometimes demonstrate that it is essentially a hostile environment, as we have been reminded by the hovercraft accident at Dover and the more terrible disaster (in terms of loss of life), the foundering of the *Herald of Free Enterprise* early in 1987 outside Zeebrugge.

Has the challenge of the Channel been met? On the main services, the speed of crossing has not improved in 100 years. There remain the problems of bad weather, and the awkward transhipment of passengers between land-mode and sea-mode of travel. Another factor is the enormous rise in the cost of replacing ships – and of course aircraft, hovercraft and jetfoils – all of which are relatively short-lived assets. The rise in replacement costs has exceeded the fall in money values – ie it outstrips inflation. That is probably the most important reason why fares on this short stage of 21 miles or so are, per mile, about the highest in the world. It is also a very strong argument for building something not subject to renewal at increasing cost every few years – namely a fixed link. That is a way to meet, once and for all, the challenge of the Channel, just as years ago the challenge of the Alps was met by driving the great Alpine tunnels that link Italy with France and Switzerland.

2 The Visionaries

It was not with the idea of facilitating the invasion of England that during the short-lived Peace at Amiens, in 1802 a French mining engineer from the Département du Nord, Albert Mathieu-Favier (usually known just as Albert Mathieu) obtained an audience of Napoleon Bonaparte in order to submit a plan for a Channel Tunnel. Napoleon was then First Consul, not yet Emperor, and some Englishmen still saw him as a champion of democracy against corrupt monarchies; Charles James Fox was among them.

Mathieu showed Bonaparte his plans. They involved boring two tunnels, one from France and one from England, rising up to an artificial island to be constructed on top of the Varne bank. This would be a posting stage, for change of horses. Mathieu also envisaged interesting developments on the Varne which would become a small city in mid-Channel. His tunnels would be lit by gas, and ventilation would be provided by chimneys rising from the sea bed to the open air.

Baron Emile d'Erlanger told the Royal Geographical Society in 1917 that Napoleon had liked the scheme sufficiently to say to the British Ambassador: 'C'est une des grandes choses que nous devrions faire ensemble'. Other accounts attribute those words to Charles James Fox. Mathieu got some encouragement from having his plans shown to the public at exhibitions in Paris, but that was virtually all that happened.

In 1803 Tessier de Mottray suggested a tunnel in the form of an immersed tube rather than a bored tube – an idea that still had its supporters 180 years later. He ignored the lack at that date of the technology needed to translate such a notion into practice.

Thereafter the Tunnel was virtually forgotten until about 1830 when A. Thomé de Gamond (1807–1876), a remarkable polymath (he was a civil hydrographic and mining engineer, a Doctor of Laws, a Doctor of Medicine, and a reserve officer in the Army Engineering Corps in France), began to be fired with enthusiasm for a fixed link. He was the first to undertake serious geological and hydrographic studies. He had a remarkable similarity to his half-French counterpart in England,

Isambard Kingdom Brunel; they shared a passion for grand designs, and persistence in the face of discouragement.

Although deeply committed to the idea of a fixed Channel link, de Gamond was ready to consider any practicable form that the link might take. Furthermore, he was determined that any project he would sponsor would be based upon research and testing, carried out as scientifically as possible, rather than presented just as a flash of inspiration.

De Gamond began by making geological and hydrographic surveys of the Channel including samples dredged up from the sea bed. His study of the strata convinced him that tunnelling was feasible. However, in 1834 his first proposal for a fixed link was, like Mottray's, a submerged tube. It was to be of iron, encased in masonry. In 1836, no one having taken up this proposal, he prepared designs for a Channel Bridge which could be arched, or flat-topped, or tubular. The following year he made a model showing moles extending eight kilometres into the sea from each coast, linked by a huge concrete floating ferry-platform, propelled by steam.

Again, no one was prepared to support such a project and de Gamond then retired to consider the question of a fixed link in more detail. He spent many hours in a boat equipped with sounding-lines, at the end of which were hooks to bring up samples from the bottom. He dived, naked, into the inhospitable waters of the Channel to a depth of 100ft on three occasions. De Gamond's courage was extraordinary. He dived without any form of breathing apparatus, apart from filling his mouth with olive oil as a sort of one-way valve, and descended rapidly by carrying heavy stones. He did however attach himself to a line which he tugged when he needed to be pulled up by the boatman who was his companion. On reaching the surface he once reported that he had been attacked by strange marine monsters, or 'malevolent fish', which local fishermen identified as conger eels.

In 1856 de Gamond produced – after flirting with the idea of a stone viaduct of 400 spans – a scheme for a 34-kilometre tunnel in masonry, large enough to take two rail tracks. These would pass under the Varne bank, which, as in Mathieu's design, would be built up to make an artificial island. The island would be an international port – no doubt enjoying the benefits of 'duty free' – and would have the picturesque name of Etoile de Varne. A shaft and staircases would enable communication to be made between ships above and tunnel trains below.

De Gamond's timing was quite good, since this was a honeymoon period in the relations between Britain and France. Allies in the

Crimean War, memories of Waterloo seemed at last to have been forgotten in both countries. Queen Victoria and Prince Albert paid a State visit to Emperor Napoleon III and Empress Eugénie, and were charmed with their reception. A Channel Tunnel seemed an excellent way of cementing the new relationship, and Napoleon directed the French Commission for Scientific Research to examine de Gamond's scheme. The Commission gave it a favourable report, but stressed that full Anglo-French co-operation at governmental level would be essential. The French Ambassador in London urged his government to assist in developing the scheme. On the British side, the Prince Consort was keenly interested and Queen Victoria, who suffered from sea-sickness, was reported to have said 'You can tell the French engineer that if he succeeds, I will give him my blessing, and in the name of all the ladies of England'. Lord Palmerston, however, who may not have ever been sea-sick, told the Prince Consort that the Prince would not be so keen on the tunnel if he had been born on an island.

Jumping on the band-wagon of a Franco-British rapprochement, other French engineers and economists put forward ideas for a fixed link. An eccentric named Horeau produced a fanciful plan for a submerged tube from which ventilating shafts would lead upwards to a series of floating Gothic pavillions decorated with turrets and pennons. A Dr Prosper Payerne had already submitted to Napoleon III in 1852 a scheme for a huge causeway as the foundation for a tunnel carrying a railway. Payerne's special claim to fame had been his invention of a diving bell, which he patented and demonstrated in the middle of Paris by himself staying submerged in the Seine for an hour, attracting a large crowd. His tunnel scheme would employ diving bells for the construction of the causeway from pre-fabricated concrete blocks on the sea-bed; walls would then be built and enclosed to form a tunnel, within which a railway would run – the whole work being carried out under air pressure.

Contrasting with de Gamond's zeal for investigation and – so far as possible – testing the practicability of the ideas his fertile brain threw up, a long succession of proposals appeared around the middle of the century which never got beyond the drawing board or in some cases even the sketchbook. Both immersed tubes and floating tubes were repeatedly suggested in schemes proposed in the 1850s (Horeau, Wytson, Nicholl, Vacherot, Angelin, Favre, Turner); around 1861 (Smith, Chalmers); and 1869 (Martin and Leguay, Marsden, Batesman and Revy, Zerah Colburn, Page, Bishop).

A Channel bridge attracted almost as many proponents, after Thomé de Gamond had abandoned the idea. About 1867 two schemes were put up (Boutet, Boyd); in 1875, another (Mottier). The bridge idea in fact led to the formation of a company to develop a plan emanating from Vérard de Saint-Anne; it did little more than engage in propaganda and was succeeded by the Channel Bridge and Railway Company of 1884, which got as far as exhibiting plans drawn up by the firm of Schneider et Cie of Le Creusot and M. Hersent in association with Sir Benjamin Baker, the engineer of the Forth Bridge.

Lastly, a bridge-plus-tunnel idea appeared as early as the 1870s, put forward by Bunau-Varrilla. It was to reappear a century later.

None of these projects led to any physical works being started, although they contributed to a strong body of opinion being formed in France in favour of a fixed link with the United Kingdom. The Duc de Persigny, an important politician, and Michel Chevalier, the economist, were notable fixed link supporters.

But it was de Gamond's persistence that made the real impact, above all on the British side. In 1866 he produced a modified version of his 1856 tunnel scheme, with some of the more fanciful elements omitted; it passed, like its predecessor, under the Varne bank but now incorporated spiral tunnels instead of spiral stairways, leading up to the artificial island which de Gamond still proposed to create.

And the British side did not remain passive. Stimulated perhaps by Royal approval, John Hawkshaw (1811–1891, who was to become Sir John in 1872) one of the most eminent consulting engineers of his generation, began to interest himself in a Channel Tunnel. He had been a mining engineer in Venezuela but returned to England to design major works for several railways. In 1861 he became consulting engineer of the South Eastern Railway, for which he designed both Charing Cross and Cannon Street stations, including the barrel roof of the Charing Cross which was to collapse in 1905. He was to be joint engineer of London's Inner Circle project, and he advised the Khedive of Egypt on the Suez Canal. His mining background, like his South Eastern Railway appointment ensured that he would have an interest in the Channel Tunnel. It was to last a long time.

However, his early ideas were based on a single-bore tunnel large enough to carry two railway tracks. William Low, (1818–1886) an engineer who had been working with him on the scheme, disagreed strongly and urged the adoption of two single-track bores. Low had important arguments on his side. The cost of tunnelling increases, not just in proportion to the diameter, but considerably more. And every increase in diameter increases the risk of emerging from the Lower Chalk stratum and entering less favourable tunnelling formations. He pressed his views to an extent that led him to part company with

Hawkshaw. Low then, in 1867, collaborated with de Gamond to produce the latter's last proposal, bringing in another British engineer named Brunlees, who had designed railway and harbour installations. The scheme was for twin railway tunnels and again the timing was good, for it more or less coincided with the Great International Exhibition in Paris – Napoleon III's answer to the Prince Consort's Great Exhibition of 1851 in London.

William Low stressed the need to solve the problem of ventilation, though he also pointed out the piston effect of the trains passing through his single-track tunnels, which would be much less with a large double-track tunnel. He proposed cross-passages at frequent intervals between the two bores to assist air movement, a system often employed in long mining galleries.

The de Gamond-Low-Brunlees project was referred to an English parliamentary committee chaired by Lord Richard Grosvenor and a commission of French engineers. There was a prospect that things might really get moving at last, but it was dashed for a time by the Franco-Prussian war of 1870 leading to the abdication of Napoleon III, and then by a controversy over Hawkshaw's single bore tunnel and Low's twin tunnels. Not that Lord Richard Grosvenor's committee was out of action. It turned itself into an Anglo-French commission which enabled it to recruit Michel Chevalier; it also invited views from a number of railway engineers.

After 1872 the new French Republic was able to turn to works of peace and diplomatic exchanges between the British and French governments began to look constructive. On the British side there was a good deal of inter-departmental consultation and procrastination which – as was several times later to be the case – the French found unreasonable and irritating. At last however the two governments moved forward together, in 1875. The era of the early visionaries was over and it was now the turn of financiers and practical engineers. Thomé de Gamond died in the following year.

Despite the activity of many other visionaries, de Gamond has the best claim to be regarded as the project's real founder. But through the long years during which he devoted so much energy to the tunnel he was active on many other grand designs – for tunnels under the Irish Sea to link England and Ireland, and Scotland and Ireland; in Scandinavia across the Great Belt, the Sound and the Little Belt; across the Strait of Messina; between Corsica and Sardinia; under the Strait of Gibraltar, the Bosporus and the Dardanelles. He was engaged in finance (a plan for a credit corporation for the fishing industry) and in politics (a plan for a federal constitution in France). He was an astonishing man.

3 Watkin versus Wolseley

In the early 1870s a shrewd outside observer – say, a well-informed citizen of the United States – might have concluded that the odds were now strongly in favour of a tunnel being built. The factors that were to disprove this were several – the unpredictability of British politics, the attitude of the naval and military establishment, and to some extent also the nature of the English railway network, which lacked the strong financial backing and government support that the Nord Railway enjoyed in France.

For a time, the two governments seemed to share enthusiasm for the tunnel. The Anglo-French commission carried on its work and eventually its activities bore fruit. On 2 August 1875, Bills were passed in the British Parliament and the French Assembly. The British measure became The British Channel Tunnel Company (Limited) Act. Both in England and in France, the exercise of all powers was to be subject to prior government approval.

Passing the legislation was quickly followed by a protocol, drafted by the Anglo-French Commission. It contained the draft of a treaty covering such matters as the frontier point in mid-Channel up to which the two national systems of law would apply, and establishing conditions which the British and French companies must meet, especially the signing of agreements to start construction within five years – which could be extended to eight on request – with a time-limit of 20 years for the tunnel to be completed. An international commission would supervise construction, operation, charges, and conditions of carriage. After completion of construction, the companies' concessions would last 99 years, after which they would revert to the governments.

In many respects the protocol laid down just what all governments were to require later on, even in 1974. The signatories were, on the British side, Lord Clarendon and Lord Derby; on the French side, the Marquis de Lavelette and the Comte de Jarnac.

Ever since 1865 Hawkshaw had been working with a syndicate he had formed on studies around St Margaret's Bay and at Ferme Mouron, 2½ miles west of Calais. His proposed tunnel was a single bore of 21ft diameter which, as mentioned, caused William Low to break away. This was the origin of the confusing situation which arose in England, where competing designs led to the formation of rival companies, each seeking to build the tunnel. Hawkshaw's syndicate formed the English Channel Company, based on the St Margaret's Bay site, with financial support from the London, Chatham & Dover Railway, while Low took his ideas to Edward (later Sir Edward) Watkin, then chairman of the South Eastern Railway. Low's ideas were largely based on his experience as a mining engineer and they included the principle of ventilating underground workings by installing cross-passages at frequent intervals between twin tunnels. His standing with Watkin was reinforced by having worked on the Great Western Railway and by his association with Robert Stephenson.

Watkin was impressed by Low and put him in touch with Francis Brady, the South Eastern's chief engineer. In March 1875 Watkin obtained board sanction to spend £20,000 on tunnel studies, though the board stipulated that this was conditional upon the LCDR putting up a similar amount. The SER tunnel activity was carried on through a subsidiary called the Anglo-French Submarine Railway Company, chaired by Watkin.

Matters were also progresing well on the French side, where the organisation was much simpler. The Nord Railway, associated with the bank of de Rothschild frères, and with a Rothschild as its chairman, sponsored in 1874 the creation of a French Channel Company, with a capital of £80,000. The company came into legal existence on 1 February 1875 and in the following August the French Government formally granted it the concession for preliminary works. The shares were held in the proportion of 50% by the Nord Railway Company, 25% by de Rothschild frères, and 25% by minority interests. It started exploratory works at Sangatte, near Calais.

Not much had been done by the Anglo-French Submarine Railway Company before 1881 when Watkin launched another SER – backed undertaking, the Submarine Continental Railway Company, which planned to start its preliminary works at Shakespeare Cliff, west of Dover. It had a capital of £250,000.

Meanwhile Hawkshaw's Channel Tunnel Company, supported by the London, Chatham & Dover Railway, continued the work at St Margaret's Bay. The company required statutory powers to acquire lands and commence work, and these had been given by the 1875 Act. Watkin's original company was still at this time some way behind; it did not apply for statutory powers until 1880. Nevertheless, Watkin soon began to dominate the scene, once his new Submarine

Continental Railway Company got going.

Watkin is best known for his dynamic chairmanships of the Manchester, Sheffield & Lincolnshire, the South Eastern, and the Metropolitan railways, and for his long-standing rivalry with James Staats Forbes who led two companies in frequent competition with Watkin's empire, the London, Chatham & Dover, and the Metropolitan District.

Watkin was born in Salford in 1819, and his home remained in the village of Northenden, south of Manchester, until his death. His native territory had some cause to be grateful to him as in early life he started the Saturday half-holiday movement in Manchester and wrote *A Plea for Public Parks*. At the age of 26 he became secretary of the Trent Valley Railway, which became part of the London & North Western and, at 34, general manager of the Manchester, Sheffield & Lincolnshire Railway, becoming its chairman eleven years later. His early idealism (though not his fondness for grandiose projects) seemed to disappear once he entered the tough, wheeling and dealing world of Victorian railway politics, of which he became a master. In 1866 the South Eastern Railway, which was suffering the consequences of past mistakes, (chiefly, the arrival of competition from the London, Chatham & Dover Railway, a line which should have been bought up years ago) invited him to become chairman. He took on the job with relish, enjoying the prospect of a tussle with James Staats Forbes who had been general manager of the LCDR since 1861 and was to become chairman in 1874.

Watkin had become a Liberal MP for Great Yarmouth in 1857; he represented Stockport from 1864 to 1868 and Hythe from 1874 to 1895, ending as a Liberal Unionist. Despite parting from Gladstone over home rule for Ireland, he retained his leader's friendship and esteem. That may indeed have influenced Watkin's decision to construct a branch line to Hawarden, where Gladstone lived.

Despite his multifarious activities Watkin remained at heart provincial, his home remaining at Northenden where he had a beautiful house called Rose Hill. His rival, Staats Forbes, was a smoother, more sophisticated personality; he lived in an elegant house on the Chelsea bank of the Thames, surrounded by the paintings of the Norwich school which he collected, and fine antiques, especially china.

Watkin was nevertheless highly skilled at lobbying and public relations. The influential *Railway Times* almost served as his mouthpiece when he desired to attack the policies of the LCDR. Watkin now applied these skills to obtaining powers, similar to those enjoyed by the St Margaret's Bay syndicate, for his own company and received these through clauses in the South Eastern Railway's bill in 1881. Watkin then told the SER shareholders: 'The dream of Thomé de Gamond has come true, thanks to our friend and colleague, William Low'.

Meanwhile Hawkshaw's Channel Tunnel Company had been working on at St Margaret's Bay. Hawkshaw had commissioned some useful studies from 1865 onward. An experienced geologist, Hartsinck Day, had been engaged to make a geological survey of the chalk formations on the English and French coasts and estimates of the composition of the sea-floor. Hawkshaw arranged for H. M. Brunel to take soundings across the Channel which settled the continuity of the upper cretaceous beds. Borings were also made to test the thickness of the Lower Chalk. By 1867 Hawkshaw had put a plan forward to the Anglo-French commission for a single-bore tunnel from the neighborhood of St Margaret's Bay to Ferme Mouron. Matters more or less rested there; Staats Forbes had not so much money as Watkin to put into the tunnel.

The French found it easier to deal with Watkin because of his commitment to the Low/de Gamond design for the tunnel. The Nord Railway moreover had always enjoyed closer relations with the SER than with the LCDR, dating back to the day when the former's Folkestone–Boulogne service was well established and Calais – despite serving the mail route – was not given such a good rail connection with Paris.

Watkin's character has been described by George Dow in *Great Central* as 'imperious, incredibly energetic, and dogged with restless ambition'. 'Masterful and capricious, talented and vain, sanguine and impetuous' was how the *Railway Times* described Watkin on his retirement in 1894. It is astonishing how much time Watkin managed to devote to the tunnel, when one considers his vast range of interests – chairman of three important railways (also being chief executive of one, the South Eastern) and chairman or director of numerous others including at times the Great Western, the Great Eastern and various overseas railways. He was for a time president of the Grand Trunk Railway of Canada.

The construction in 1878 of a short spur linking the LCDR and the SER at Metropolitan Junction, on the SER's Charing Cross line between Waterloo Junction (as it was then called, now Waterloo Eastern) and London Bridge was probably an element in Watkin's grand design of running through trains from Manchester to Paris, using the Manchester, Sheffield & Lincolnshire and Great Northern Railways to reach the Widened Lines of the Metropolitan Railway, and then using running powers over the LCDR Blackfriars railway bridge and on to the SER via London Bridge to the Channel Tunnel. Later, when the MS&L had been renamed the Great Central and reached London via Aylesbury and the Metropolitan Railway, in theory at any rate it could have been envisaged that passengers in through trains to Paris would traverse the smoky Inner Circle between Baker Street and Farringdon Street! (There were of course more between sensible routes between Manchester and Dover, though these would not be exclusively over Watkin lines.)

Matters had begun moving on the French side in 1878, once the company had its concession, and some 2000yd of exploratory tunnel was bored from the Sangatte heading. Both on the French and the British side, the best method of tunnelling had been anxiously considered, in view of the advantages offered by the nature of the Lower Chalk. A boring machine rather than the traditional mining procedure of drilling, blasting, and spoil removal was indicated. Blasting in the Lower Chalk might in any case cause disturbance far outside the area aimed at, which could be dangerous under the sea, should a fault or fissure exist.

The first tunnelling was with a boring machine known as the Brunton tunneller. It had three or more revolving disc cutters, turning on their own axes and also, together, on the axis of the tunnel. It was carried on a rail track which was constantly lengthened as the cutters worked forward. Unfortunately its performance was unsatisfactory and a superior type of 'mole' was used, designed by Colonel Beaumont of the Royal Engineers, who had worked on the fortifications of Dover Castle. The Beaumont tunneller was put to work both in England and France, cutting a tunnel seven feet in diameter, though the French version, built by the Société des Batignolles, incorporated some modifications. Its operations were supervised by an experienced engineer named Welker, who had been in charge of the compressed air machinery used in constructing the Gotthard Tunnel in Switzerland. The two Beaumont machines employed the compressed air technique to advantage, being reputed to work fast, cutting at a rate of 40ft in seventeen hours – the remaining seven hours of the 24 presumably being devoted to maintenance and spoil removal. The whole looked rather like a World War I tank, the propelling machinery being encased in the body and the cutter being a rotating shaft carrying a cross-piece with cutting edges the full diameter of the tunnel. It was worked by compressed air supplied by steam pumps outside the tunnel. Forward movement was by a hydraulic ram pressing the cutter forward as the chalk was cut away. An endless chain of buckets carried the spoil to narrow-gauge rail trucks at the rear.

Watkin's company had by the spring of 1882 excavated a shaft about 160ft deep at the foot of the Shakespeare Cliff, from the bottom of which headings were driven for a total of just over 2000yd, in two directions. On the French side, the company sank a shaft 55 metres deep and from it bored a gallery 1,839 metres long. It reported that water seepage through the chalk was one litre a minute per metre of tunnel, which could easily be coped with.

The question now was, how to obtain powers to go ahead with full-diameter tunnels. No problem was likely to arise on the French side, but obviously all decisions must be taken in step; and the British Government's attitude was therefore crucial.

The euphoria in England during the early 1870s had evaporated just when the practicability of constructing the tunnel seemed to have been convincingly demonstrated. Queen Victoria told Disraeli in 1875 that she now considered the tunnel project 'very objectionable'. Had the Prince Consort still lived, he would almost certainly have taken the opposite view, as he had done in the 1850s. He would not have been swayed by emotional arguments of the kind that often afflicted consideration of the tunnel. Astonishingly, the literary, scientific and artistic world joined in opposition, in a letter to the *Nineteenth Century* signed by five national figures. Letters to the press and even leading articles began to put forward objections, sometimes highly fanciful, to the tunnel. The objectors included Thomas Huxley, Robert Browning, Herbert Spencer, and Tennyson, the Poet Laureate, who had written earlier in *Locksley Hall* about the blessings of the railway: 'Let the great world spin for ever down the ringing grooves of change' – a progressive sentiment which unfortunately showed Tennyson's mistaken belief that the flanges of a train's wheels ran in grooves!

These hidebound objections occasionally recalled the violent opposition, fifty years earlier, to the construction of the early railways. The wilder fears expressed in the 1830s – that cows would cease to give milk, and pregnant women would miscarry, at the sight of a railway train – have been laughed at ever since. But whereas the railways were built after the more extravagant fears had been shown to be groundless, this time the effect was different. Humphrey Slater and Corelli Barnett have written: 'The major fear of the objectors was of hordes of armed Frenchmen pouring through the tunnel and driving on to London with all the well-known élan that had so disturbed the Duke of York eighty years before. But the ingenuity of hysteria invented more varied objections. There was, for example, the facility that an easy means of communication with the continent would give to smugglers of small objects; the consequent inconvenience to travellers of *having their persons and baggage searched*! A more involved objection was that the tunnel would make it easier for the revolutionary societies that were supposed to teem in every continental country, and especially France, to work with cognate associations in England'.

Scaremongering about the tunnel even took the form of sensational fiction. An author calling himself 'Grip' (use of a pseudonym suggests that he may have been commissioned by tunnel opponents) produced a 'shilling shocker' – that was in fact its price – published in 1882 by Sampson Low, Marston and Co, entitled *How John Bull lost London or, the Capture of the Channel Tunnel.* It ran to 127 pages and graphically described an England invaded by French soldiers shortly after the tunnel had had a ceremonial opening. This was to be done by French 'tourists' arriving at Dover in numbers and in the dead of night seizing the tunnel entrance so as to allow an invading army to advance on London. The capital fell easily but eventually British forces moving from the provinces managed to contain the enemy and push them back to the tunnel, through which they escaped, the tunnel then being destroyed by the British!

Against this, and more serious forms of propaganda, Watkin was well equipped to cope, though he had to cajole rather than bully the government, parliament and the press; the tough tactics he used in the railway world were out of place here. His main concern was to demonstrate how easily the tunnel could be built; to prove this, visits to the workings were important. The Prince and Princess of Wales came down to inspect; and in March 1882 Gladstone, then Prime Minister, accompanied by Mrs Gladstone, Lord Salisbury and others, visited the site. Two days later an even more crucial visit of inspection was made by the Commander-in-Chief of the Army, the Duke of Cambridge, and Sir Garnet (later Lord) Wolseley who was commandant of Dover Castle. Gladstone was, in principle, in favour of the tunnel idea: the Duke and Sir Garnet took the opposite view, with some vehemence.

Watkin did not omit to court the press. Rather spectacularly, he equipped the tunnel workings with electric light (thought it was only switched on for visitors) provided by a small steam engine and generator on the surface. He decorated part of the tunnel with potted palms and the necessary facilties for serving champagne and refreshments. Special trains from London were arranged, and receptions or luncheons in the Lord Warden Hotel in Dover were provided, for instance for a press trip on 21 February 1882. The visitors were impressed by the compressed-air locomotives that ran along the narrow-gauge track to remove the spoil, and the three steam engines that drove the compressors for the Beaumont machine.

The military establishment was however becoming uneasy. On 23 February 1882 a Channel Tunnel Defence Committee was set up by the War Office chaired by Major-General Sir Archibald Alison, the army intelligence chief, to examine 'the practicability of closing effectually a submarine railway tunnel' in case of war. A fortnight later the Board of Trade warned Watkin that the foreshore below high-water mark was Crown property and, by implication, he must not trespass upon it without parliamentary authority. Watkin returned a typical reply, namely that his company had acquired by purchase of land certain ancient manorial rights originally granted by the Crown covering the foreshore. On 1 April the Board of Trade dismissed this argument, reiterating that below low water mark the sea-bed was Crown property quite irrespective of any rights over the foreshore, and it instructed Sir Edward to suspend work immediately pending a report from the Alison committee.

Watkin's tactics were predictably devious. He pleaded that if the work stopped completely the ventilation (provided by the compressed-air plant) would become so defective that it would endanger the lives of workmen still in the tunnel. He obtained a short respite on this ground, but the Board of Trade, rather suspicious, said that its chief inspecting officer of railways, Colonel Yolland, would come down to investigate. Watkin stalled for several weeks, refusing to agree a date for the inspection. Finally in June the Board lost patience and obtained an injunction from the High Court restraining the company from any further tunnelling work without the consent of the Board. It was 15 July before Colonel Yolland could make his long-deferred inspection; he was convinced that tunnelling *had* been going on surreptitiously. Watkin, under threat of committal for contempt of court, finally gave up; he announced this at one of his tunnel parties which included Ferdinand de Lesseps, the builder of the Suez Canal. He said that he had just learnt that he would have to appear before a court of law for having committed 'the crime of carrying on these experiments' – news greeted with hisses and groans by the company present!

Meanwhile with military promptitude if not sagacity, the Alison committee had reported as early as 12 May that the tunnel was a potential danger to national security. The government's next step was to appoint a joint select committee of both houses of parliament, under the chairmanship of Lord Lansdowne, to examine all aspects of the tunnel project. The committee started its work quickly, but did not report as precipitately as the Alison committee; it met, usually twice weekly, for 14 months and studied 574 pages of evidence.

The controversy raged while the works were suspended. The most decisive contribution is generally considered to be Lieutenant-General Sir Garnet Wolseley's memorandum to the joint select committee. This is a surprising document, since the adjutant-general did not deal with the military aspects as a matter of logistics but indulged in rhetoric, for instance asking if England shall cease to be a 'sea-girt isle'; arguing that 'soldiers and sailors view the scheme with horror and undisguised alarm' (untrue as a generalisation, as is shown below); and, 'fearing that another Napoleon may arise', suggests that 'surely, John Bull will not endanger his birthright'.

It is not necessary to describe at length all military arguments for and against the tunnel that raged in the 1880s and for a surprising length of time afterwards. But it is astonishing to read the Wolseley memorandum, in view of the decisive effect it seems to have had in influencing the joint select committee. It is a remarkably inept production from one holding nearly the highest military rank. It discusses matters with which Sir Garnet was unqualified to deal – the economic and financial case for the tunnel – yet is woolly and in fact incorrect in dealing with military logistics. The least one could expect would be a careful appraisal of the tunnel's through-put capacity for an expeditionary force, covering entraining and detraining facilities and times, rolling stock requirements, facilities for working back empty stock, turning, watering and possibly coaling locomotives and so on. None of this is touched on. There is simply a wild guess that 20,000 men could come through the tunnel in four hours.

The French general staff, who later on made a careful study of tunnel military logistics, concluded that 48 and not four hours was nearer the truth: Wolseley had forgotten or purposely omitted the requirements of artillery, stores, equipment, road transport, victualling and all the support services including ambulance trains, ammunition trains and so on. The French general staff later on calculated that *with the co-operation of the English authorities,* an expeditionary force of 150,000 men could be transported from England to France, or vice versa, in fifteen days, provided full preparation had been made at the entraining and detraining stations.

Wolseley forecast that the construction of the tunnel would be such a danger that Britain would require to introduce conscription. 'Are we' he asked 'deliberately to make England less safe in order that tourists may not suffer [from sea-sickness]?' The Duke of Cambridge supported Wolseley's views in a memorandum for the Cabinet. The navy, in the person of Admiral Lord Dunsany, added its voice. His main argument was that the British navy was not as strong as it should be and that Dover Castle was not a reliable fortress – he described its guns as 'generally speaking, of an obsolete pattern – popguns in fact'.

Colonel Beaumont, who had worked on the fortifications of Dover Castle for three years, presumably to the satisfaction of the commandant, Sir Garnet, disagreed. He wrote a long and well reasoned article in the *Nineteenth Century* arguing that a whole string of devices could be employed to render the tunnel useless for the purposes of invasion by a surprise attack, any one of which would be fatal to the enemy. The tunnel entrance would be under the command of a battery of artillery; there would be secret arrangements for flooding the tunnel; and the tunnel works would be made vulnerable to fire from the British navy, even if they should have been captured.

While the select committee was at work the two surviving companies – the Hawkshaw company of St Margaret's Bay and Watkin's Submarine Continental Company – both deposited bills for authority to proceed with the tunnel. These were referred to the joint select committee. The result was a majority report to the House of Commons against the bills though it was noteworthy that of the ten members four, including the chairman, disagreed and thought the bills should proceed. This illustrates how close political decisions were, and were to continue to be, about the tunnel. A handful of votes could change the course of history.

The French company, dismayed by the British Government's decision, stopped work at Sangatte on 18 March 1883. On 1 July 1883 Watkin finally closed down the works at Shakespeare Cliff. Gladstone remained (in principle) favourable to the tunnel but he felt that the weight of hostile opinion was too great and that the Board of Trade's action in stopping further work must have Cabinet approval.

Watkin refused to admit that the battle was lost for good. His first move was to clear away the confusion caused by the continuing existence of two tunnel companies in Britain. The London, Chatham & Dover Railway, always hard up, was quite glad to dispose of its interest in the company which still owned the Hawkshaw explorations at St Margaret's Bay. In 1886 Watkin increased the capital of his company to £275,000 and bought up the St Margaret's Bay organisation; a year later he changed the title of his company to Channel Tunnel Company, under which name it was to continue in existence for some 90 years.

By 1890 Watkin was ready to resume his campaign for the tunnel. On

5 June of that year he made a long speech in the House of Commons. By that time, of course, Gladstone's Liberal administration had been replaced by a Conservative government, the Prime Minister being Lord Salisbury and the key position of President of the Board of Trade being held by Sir Michael Hicks-Beach. In his speech Watkin made play with the fact that the protocol of 1876 had been signed on behalf of a Conservative government headed by Disraeli. Hicks-Beach made a stonewalling reply, repeating the military objections to the tunnel; but Gladstone from the Opposition forcefully supported Watkin's arguments. He attacked the military objections and strongly emphasised the economic case for the tunnel. However, the House refused permission to restart work on the tunnel by a majority of 81. It was to be 1906 before parliament was to be asked once more to look at the Channel Tunnel and by that time Watkin was no longer there to play a part. In 1894 his health failed to an extent that required him to give up his main preoccupation, the Manchester, Sheffield & Lincolnshire Railway chairmanship, as well as those of the South Eastern Railway, the Metropolitan Railway and lesser undertakings including the Channel Tunnel Company. He died in 1901.

As a postscript to Watkin's Channel Tunnel activity, it should be recorded that after 1890 he turned his attention to a seam of coal which the sinking of the shaft at Shakespeare Cliff had discovered, and a small colliery was established beside the Channel Tunnel trial works. It was not successful, but it settled something that had long been suspected, that East Kent contained a coalfield. Francis Brady, the SER chief engineer, was active in developing new mines, though, rather ironically, the seams that proved worth exploiting commercially were in the territory of Watkin's enemy, James Staats Forbes; they were opened up after Watkin's death in the Betteshanger, Tilmanstone, Snowdown and Chislet collieries, all situated much closer to the former London, Chatham & Dover main line than to that of the South Eastern.

4 The Early Twentieth Century

It would be wrong to think that after Watkin's death the tunnel ever lacked champions for long. From 1900 until the first world war things moved again in favour of the tunnel, stimulated by two developments. The first was the change in Anglo-French political relations. These had deteriorated in the 1880s and 1890s, and had been at their lowest point in 1898 when the Fashoda incident – a French military incursion into the Upper Nile area which aroused British anger and suspicion – was followed by the Boer War which made the British generally unpopular on the Continent. From then on matters steadily improved. King Edward VII was a Francophile and in 1903 he paid a state visit to Paris. In 1904 an Anglo-French protocol was signed – the so-called 'Entente Cordiale' – that grew into an alliance designed to counter the growing threat of German militarism spearheaded by the ambition of Kaiser Wilhelm II. The logic of providing a fixed link between allies seemed strong.

The second factor was the growth of electric traction technology, which greatly eased the problem of ventilating the tunnel, a problem that had probably been dismissed rather too lightly when steam traction through a tunnel 34 kilometres long without ventilating shafts from the surface had been contemplated, though Colonel Beaumont had rather optimistically suggested compressed-air locomotives. Admittedly the great Alpine tunnels had no ventilating shafts and the Gotthard and Mont Cenis tunnels were used by steam trains. But ventilation was a constant headache in Britain in certain tunnels, especially the Severn and Woodhead bores, respectively just over four and three miles long; the Channel Tunnel would be more than twice as long as the longest Alpine tunnel. Now, however, main line electric traction had been introduced in Europe and in the USA and obviously would be employed in any major tunnel. It would not only reduce ventilation problems but its capacity to cope with rising gradients better than steam would simplify the problem of the vertical profile required to ensure that the tunnel's axis remained within the Lower Chalk stratum.

New technical studies of the tunnel were undertaken by Albert

Sartiaux of the Nord Railway and Sir Francis Fox, while Percy Tempest, chief engineer of the South Eastern & Chatham Railway continued the studies of his predecessor, Francis Brady, and also acted as a technical adviser to the Channel Tunnel Company. Sartiaux was manager of the Nord and also chief government engineer for roads and bridges – a combination of responsibilities that could not exist in Britain. Sir Francis Fox was a leading consulting civil engineer of his day and founder of the firm of Freeman Fox and Co. These two distinguished men were also advisers for the Simplon Tunnel project.

They agreed that the de Gamond/Low principle of twin single-line tunnels was correct but they revised the proposed alignment to keep more securely within the Lower Chalk; they paid more attention to the technique of spoil disposal and accepted that even with electric traction, ventilation needed more provision than previous planners had thought necessary.

Their proposals also covered much more thorough treatment of the drainage that would be required. They proposed separate drainage tunnels leaving the alignment of the main tunnels from the two low points of a much elongated 'W' alignment (one which the slope of the chalk beds favoured) and leading to 'sumps' on the coast north of Dover and at Sangatte. Here there would be pumping stations to dispose of the water.

Electric narrow-gauge railways would be laid from the working faces for spoil removal – an improvement on Beaumont's compressed-air locomotives. Using the drainage galleries, cross-passages to the axes of the main tunnels would allow additional working faces for the latter to be established, greatly speeding the rate of construction.

Sartiaux and Fox had the great advantage that much tunnelling experience had been gained recently, not only in the Alps but from the construction of the London Underground railways through heavy clay, using Greathead shields at the working face.

With the political background apparently favourable and with improved technical planning, the prospects once more looked good. Baron Emile d'Erlanger of the great international banking family, second only to the Rothschilds, was now chairman of the Channel Tunnel Company. On the French side the Association du Chemin de Fer Sousmarin entre la France et l'Angleterre had enjoyed Government backing and financial support exceeding anything the privately-owned railways in Britain had felt able to give their own tunnel companies.

Accordingly in December 1906 a new Channel Tunnel Bill was introduced into the House of Commons where a Liberal Government had recently arrived under Sir Henry Campbell-Bannerman; in view of

Gladstone's former support for the tunnel, the bill should have enjoyed a clear passage. The bill's supporters published a number of arguments in its favour, drawn from a report made by an independent expert. This showed estimated total annual cross-channel passengers to be:

1906–1910	1.4 millions
1911–1915	1.65 ,,
1916–1920	2.15 ,,

In the first year after opening the tunnel receipts were estimated at £1,542,000 against expenses of only £375,000. Allowing for bond interest amounting to £400,000, the net receipts would allow of a divident of 7¼ per cent being paid on the ordinary stock.

Despite this optimistic forecast, the sour face of British reaction showed itself once again. A chorus of criticism and objection arose, largely repeating the outworn military pontifications of 1882 and 1883. The emotional arguments even included raising the bogey of conscription, of endangering British ability to pay old age pensions, and so on. Echoing the Duke of Cambridge and Lord Wolseley, Major-General Sir Frederick Maurice attacked the tunnel at one end of the spectrum, as did the tiny Labour Party in the House of Commons at the other. But the military view was not unanimous: two generals, Sir William Butler and Sir Alfred Turner, supported the tunnel, as did Vice-Admiral Sir Charles Campbell for the navy. However the War Office declared its opposition, and in parliament the government spokesman in both the Commons and the Lords recommended rejection of the bill. It was accordingly withdrawn by its sponsors on 26 April 1907.

It was to be six years before another move was made by those who believed in the tunnel. In that last wonderful summer of 1913, with Edwardian peace and prosperity in England, on 11 June a Channel Tunnel parliamentary committee was formed with Arthur Fell, MP, (later Sir Arthur), a Conservative, as chairman. This committee was to continue in being for many years. It included a splendid champion of the tunnel in Sir William Bull, MP, who succeeded Fell in the chairmanship after the 1914–18 war. Sir William was senior partner in a leading firm of solicitors but he found time for a very wide range of political and other activities. He is described in the catalogue of his papers, now deposited in the Archive Centre of Churchill College, Cambridge, as 'an immensely likeable man – intelligent, observant, hard-working, enthusiastic, generous and very warm-hearted'. His long-standing interest in the Channel Tunnel may have been stimulated by his experience of seeing the Blackwell Tunnel built while he was chairman of the Bridges Committee of the London County Council, though it had been sparked off by his association with Fell. He was known to close friends as Paul (and occasionally addressed as 'Taurus').

He was Conservative MP for Hammersmith (later Hammersmith South) from 1900 to 1929. He was knighted in 1905, became a Privy Councillor in 1918, and received a baronetcy in 1922. Sir William Bull liked to stress that believing in the tunnel was entirely compatible with a strong sense of British patriotism and individuality; he was fond of introducing himself with the words: 'My name is Bull and I come from Hammersmith'. As a book by one of his sons (*Bulls in the Meadow*) recalls, he once went further when visiting France and announced to a surprised customs official: 'Je suis Monsieur Boeuf, un anglais', magnificently ignoring the fact that *taureau* is the French for bull, and that no one would ever take him for anything but English! (His French was spoken with the sort of accent later associated with Winston Churchill.)

On 5 August 1913 Arthur Fell led a deputation to the Prime Minister, H. H. Asquith. The parliamentary delegation of fifteen claimed the backing of 90 MPs drawn from all the political parties. Fell handed Asquith a memorandum arguing that the only real opposition to the tunnel was still based on the Wolseley paper to the joint select committee in 1883 and that 'even at that time the fears expressed by Lord Wolseley were not shared by the military committee'. Characteristically, Asquith hedged; he was cool if not actively hostile, pointing out that this scheme has been 'opposed and resolutely opposed by every Government for 25 years'. He ran true to form by saying that more time was needed for consideration. History in fact has always associated Asquith with his catch-phrase 'wait and see'.

While waiting, the controversy built up once more. A congress of British Chambers of Commerce held (surprisingly) in Antwerp passed a motion of overwhelming support. But Admiral Sir William Horsey produced a little pamphlet *Natural Defence v Channel Tunnel,* reproducing the text of letters he had written to the *Morning Post* in 1882 and 1883; one such letter demanded 'Is it conceivable that Nelson or the Duke of Wellington would have tolerated the idea of a Channel Tunnel?' One can hardly envisage a better justification for the sneer that the service mind in peacetime is usually preparing to fight the last war – or the one before that!

The Admiral did not mention that the Duke of Wellington had

opposed the construction of a railway between London and Portsmouth on the ground that it would facilitate a French invasion. Almost tearfully Horsey ended his pamphlet by assuring his readers 'that the stupendous folly of connecting Great Britain with the Continent may never be undertaken is the prayer of the oldest Admiral in the British Navy'.

Asquith's delaying tactics did not appeal to the French Premier, Louis Barthon, who was a strong tunnel supporter, though they must have cheered the oldest Admiral in the British Navy. On 23 September 1913 the Franco-British Travel Union held a congress, presided over by Baron Emile d'Erlanger, at which Percy Tempest showed a collection of tunnel drawings, including a proposed spiral tunnel leaving the ex-SER main line in the outskirts of Dover and descending to link up with the Channel Tunnel proper, with another connection from the ex-LCDR main line, thus providing two separate rail routes from London to the tunnel.

Asquith however, as Humphrey Slater and Corelli Barnett remark 'meanwhile handed the matter over to the traditional wrecking crew; the Admiralty, the War Office and the Board of Trade'. They were to report individually to the Committee of Imperial Defence.

While the reports were being drafted, submitted and considered, Fell and his parliamentary group continued their propaganda, including a visit by VIPs to the working sites in England and France in December 1913. Not for them however, the champagne, potted palms and specially-switched-on electricity of Sir Edward Watkin: the total cost for the visit (five persons) was 95 francs 10 centimes (at that time less than £5)!

The Committee of Imperial Defence reported on 15 July 1914 against the Channel Tunnel, though its secretary, Lord Sydenham, was in favour. Less than three weeks later, war with Germany was declared. Marshal Foch was later to say that if the tunnel had been open in 1914, the war would have ended two years earlier than 1918. Certainly the tunnel would have immensely assisted the despatch of the British Expeditionary Force, and the maintenance of its supply lines thereafter, as compared with the long sea crossing from Southampton and the more hazardous if shorter routes across the Dover Strait.

During the war, 126 railway-owned cross-Channel ships were commandeered, mostly for use as military transports. Of these no less than 31 were lost by enemy action. To supplement cross-Channel transport facilities, a rather primitive train ferry service was started in 1918, based on Richborough in Kent. The need for a tunnel must have stared the military in the face.

Arthur Fell and his colleagues tried to ram the message home. They put down no less than five parliamentary motions for debate between 1914 and 1922 when the coalition government headed by Lloyd George was replaced by a Conservative administration. In 1917 the chancellor of the exchequer, Bonar Law, actually told Fell that it was not 'opportune' to discuss the tunnel while the war continued. From 1919 to 1922 no less than nineteen attempts, other than by putting down motions, were made to get parliament to reconsider the whole question. Despite the malaise that crept over British politics in the inter-war years, the supporters of the tunnel refused to be silenced – and they were to acquire a notable recruit in Winston Churchill.

5 The Inter-War Years – and Churchill

The experience of 1914–18 would seem to have made obvious the military desirability of the tunnel to both Britain and France. So at any rate it appeared to the French. They had never envisaged the tunnel as a military threat to their own country; on the contrary, the lack of it during the four grim years of war with Germany had been a serious handicap. The French had conceded that British fears might have to be assuaged, and they had offered more than once to plan the approach railway on their side to run in an exposed position in which it could be destroyed by gunfire from the British Navy if hostilities with Britain were ever to break out. More recently the Sartiaux-Fox plans had included new safeguards. The elongated 'W' vertical axis meant that there would be two low points, and either or both of these could easily be flooded by remote controls, operated from either country.

The French view of the strategic need for the tunnel by both Britain and France had been echoed in a prophetic article by the creator of Sherlock Holmes, in the *Graphic* magazine in February 1913. Conan Doyle had then correctly forecast the 1914 situation: he wrote 'If France be wantonly attacked, we must strain every nerve to prevent her going down, and among the measures to that end will be the sending of a British expeditionary force to cover the left or Berlin wing of the French defences. Such a force would be conveyed across the Channel in perhaps a hundred troopships and would entail a constant service of transports afterwards to carry the requirements.' Conan Doyle went on: 'We are now bound in close ties of friendship and mutual interest to France. We have no right to assume that we shall always remain on as close a footing, but as our common peril [the threat from Germany] seems likely to be a permanent one, it is improbable that there will be any speedy or sudden change in our relations. At the same time, in a matter so vital as our hold upon the Dover end of the tunnel, we could not be too stringent in our precautions. The tunnel in fact should open out at a point where guns command it, the mouth of it should be within the lines of an entrenched camp, and a considerable garrison should be kept permanently within call. . . . As an additional precaution, a passage should be driven alongside the tunnel, from which it could, if necessary, be destroyed.'

The French tried to bring home the lesson of 1914–18 almost immediately after the Armistice had been signed. A study by an engineer named Tobiansky d'Altoff claimed that post-war reconstruction must involve the digging of the tunnel to prevent any repetition of the events of August 1914. On 1 July 1919 the French government received a report from a ministerial committee headed by M. Colson pointing out that the French 'Association' was still in being and that under the 1875 protocol it continued to hold the exclusive right of construction and exploitation, so far as the French share in the tunnel was concerned. The lack of progress had been due solely to British governmental obstruction.

The British Channel Tunnel Company had already reopened its studies; it published a new report on the technical feasibility of the tunnel on the same day, 1 July 1919, as the Colson committee confirmed the standing of the French company. A French engineer named Fougerolles now suggested a pilot gallery; this was in fact the ancestor of the service tunnel that has been included in the 1973 and subsequent tunnel plans. Fougerolles also suggested an ingenious method of spoil disposal which was still under consideration as late as 1965–70 – mixing the excavated material with water and pumping it up through shafts piercing the sea floor, where it would emerge under pressure in the form of a slurry.

Another interesting technical development, this time from the British side, was the invention of a new boring machine, more effective than the Beaumont 'mole' of 1882. It was designed by a Leicester engineer, D. Whitaker, and originally had been submitted to the Royal Engineers for use in tunnelling beneath the enemy lines during the war. Percy Tempest bought an example of the Whitaker boring machine for the Channel Tunnel Company in October 1919 at a cost of £6,150. It was tested in the Warren, between Folkestone and Dover, where it excavated at a rate of 40ft in 10 hours. It was operated by an electric motor of 120 horse power. After the trials it remained for many years embedded in the chalk cliff; at one point a scrap merchant made an offer to buy it at the derisory price of £6, which was deservedly rejected.

The post-war depression in Britain's heavy industries appeared soon after the brief boom of 1920 had collapsed. Consequently the Channel Tunnel parliamentary committee listened with interest to Sir Arthur Fell when he drew attention to the need to relieve unemployment by public works such as the tunnel. Again, he was unwittingly prophetic – in little more than a decade, Hitler would relieve German unemployment with the construction of the autobahn motorway network while Britain still tinkered with class A roads of obsolete alignment and standards of construction.

In 1921 the Franco-British Association formed a French Channel Tunnel committee parallelling Fell's British committee, with Marshal Foch as honorary president; this was followed by a Belgian Channel Tunnel committee. Unfortunately, bold ideas or constructive thinking seemed foreign to all the governments that were to preside over a sadly deteriorating Britain for the greater part of the inter-war years. In 1922 Sir Arthur Fell handed over the chairmanship of the parliamentary committee to Sir William Bull who with 'bull-dog' determination (as he himself liked to suggest) was to continue fighting for the tunnel until his death in 1931.

The first, short-lived, Labour Government took office in 1924 and Ramsay Macdonald, the Premier, was immediately approached by Sir William with a request that the government should re-examine the tunnel. Macdonald felt he must show that he was not entirely obstructive – he had in fact while in Opposition made favourable references to the tunnel. He then however took the curious step of calling a meeting under his chairmanship of four former Prime Ministers, Asquith, Lloyd George, Balfour, and Stanley Baldwin. The meeting lasted only 40 minutes; each of the former Premiers recalled that there had been government hostility to the tunnel during his time in office, and saw no reason to reopen the matter.

Macdonald felt comforted by this powerful support for a do-nothing policy — always dear to his heart – and on 16 June 1924 he rose to make a statement in the House of Commons. He agreed that previous reports were now out of date and that the whole question needed to be re-examined. Having appeared to give a smile of welcome, Macdonald then administered the kiss of death – reference of the whole question to the Committee of Imperial Defence. Within three weeks that body ran true to form and restated its opposition to the tunnel.

Some years later on 8 January 1929, Bull wrote to Emile d'Erlanger 'In 1924 Ramsay Macdonald was strongly in favour of the tunnel but as he said to me afterwards, "What can I do? As a constitutional Prime Minister I am bound to accept the verdict of the Imperial Defence Committee".'

In the same month of July 1924 the tunnel's most illustrious defender, Winston Churchill, entered the debate. In an article in the *Weekly Dispatch* he castigated the ex-Premiers who had got Macdonald off the hook. 'In forty minutes five ex- or future ex-Prime Ministers dismissed with an imperial gesture the important and complicated scheme of a Channel Tunnel for which the support of four hundred members of Parliament had, it is stated, been obtained. Four hundred members. Five Prime and ex-Prime Ministers and forty minutes. Quite a record! One spasm of mental concentration enabled these five super-men, who have spent their lives in proving each other incapable and misguided on every other subject, to arrive at a unanimous conclusion.'

Churchill's biting sarcasm continued: 'They had the advice of the united general staffs of the navy, the army and the air force. So clear, so cogent, so convincing were the arguments which the professional heads of the three fighting services presented during that period of forty minutes reserved to them, that discussion was superfluous or impossible. . . . The public have a right to know what were the grounds on which a great decision like this was taken.'

Churchill was in good form when he touched on the service chiefs' mentality. 'We know how the Admiralty resisted the introduction of steam in the Royal Navy, derided the propeller, and discounted the submarine. The great Duke of Wellington delivered a speech in the House of Lords proving conclusively that a locomotive engine could not draw any more than its own weight along a track of rails unless provided with a rack and pinion'. He recalled derisively that the military and naval authorities had resisted the landing of a submarine cable on the shores of England for fear that this would assist an invasion; and that they had objected to the Great Exhibition of 1851 for fear that so great 'a congregation of foreigners in London' might lead to 'an attempt on their part to seize our Island home'.

Winston Churchill, whose filial piety had its limits, went on in the same article to recall that his father, Lord Randolph Churchill, who had opposed the tunnel, had used the phrase 'the reputation of England has hitherto depended upon her being, as it were *virgo intacta,* and had argued that all ideas of closing the tunnel in wartime were illusory. He had convulsed the House of Commons with laughter in the 1880s by pointing to the ministers on the Treasury Bench and exclaiming 'Imagine the present Cabinet sitting around the table and debating who is to press the button and when!'

Nothing more happened for years in Britain, although Sir William Bull kept tabling motions for a debate in the House of Commons. In November 1925 the Prime Minister, then Baldwin, refused to re-open the subject; in November 1926 he repeated his stone-walling; in May 1927, again he 'had nothing to add'. In 1928 he invoked the well-known political doctrine of unripe time, when pressed to do at least *something* about the tunnel.

Finally, in January 1929 Sir William Bull was successful in getting a debate in the Commons, and at the end of it Baldwin gave way, to the extent of agreeing that the government would remit the whole question to yet another Royal Commission. He, like the Cabinet, had been at last impressed by the energetic canvassing inside and outside parliament by Bull's committee, which suggested the existence of majorities in favour of the tunnel in both Houses, and also outside support among the trade unions, the Federation of British Industries and local authorities.

The Royal Commission sat from February 1929 to February 1930 when it reported. It delegated the detailed work of study to a sub-committee of the government's committee of civil research and it not merely instructed the sub-committee to consider the tunnel but also any 'other new form of cross-Channel communication'. This widening of the terms of reference naturally opened the door once again to the promoters of alternative schemes including some fantasies that recalled the wilder projects of the nineteenth century. The sub-committee had to waste some of its time considering a proposed jetty from Dover to Calais – a massive dam, in effect. A bridge was again proposed with spans wider and higher than those of the Forth Bridge. Immersed tubes reared their heads (to mix a metaphor) once again in more than one form.

A project that received considerable publicity was proposed in a book by William Collard entitled *London and Paris Railway.* Collard's scheme involved not merely a tunnel but also a high-speed railway on an entirely new alignment between London and Paris, using a broad gauge of 7ft (the ghost of Brunel must have smiled). He claimed a journey time, with high-speed electric trains, of 2¾ hours from London to Paris.

Collard's proposal – apart from the use of the broad gauge, which unnecessarily increased the cost to an impossible extent – was not perhaps quite as fanciful as the sub-committee thought. Only half a century later new high-speed railways in France and Japan (the *trains à grande vitesse* or TGV, and the *shinkansen* or 'bullet trains') were to demonstrate that such railways can be a financial success despite their high initial cost. Apart from the broad gauge, Collard's proposal was not so dissimilar to the one put forward and accepted by the two governments in 1973 for high-speed rail connections on each side of the tunnel. But in 1930 the sub-committee described Collard's ideas as 'impracticable in the present state of engineering knowledge and experience'.

Interestingly, Collard's proposal received support from a former deputy general manager of the North Eastern Railway, Philip Burtt, with whom considerable correspondence took place. Burtt had retired early from the railway and was lecturing on transport at the London School of Economics at this time.

However, the sub-committee's unfavourable view had been expressed even more emphatically by Baron d'Erlanger, who had written to Collard on 13 December 1928 that 'any scheme attempted outside the Southern Railway and especially the Northern of France Railway is doomed to absolute failure from the start. Both the Northern of France Railway and the French Channel Tunnel Company have charters from the French Government . . . no act of the French Chamber of Deputies and of the Senate could be obtained for a new charter to run from Paris to the coast'.

Rather confusingly, the sub-committee at one stage had its title changed to the Channel Tunnel Committee of the Economic Advisory Council. It also became known as the Peacock Committee, after its chairman, Sir Edward Peacock, a distinguished figure in both city and government circles. Canadian in origin, he had become a merchant banker – partner in Baring Brothers – and a director of the Bank of England, as well as serving on numerous government bodies.

The Peacock committee concentrated its attention upon two projects; one was the tunnel as put forward by the British and French companies and virtually following the plans of Sartiaux and Fox; the other, a train ferry. The Southern Railway and the Nord Railway were asked to estimate the effects of these alternatives upon the levels of cross-Channel traffic. Views were sought from probable critics such as the Dover Harbour Board, the shipping companies, the National Farmers Union, and the Ministry of Agriculture and the Air Ministry were consulted.

The sub-committee was not impressed by the proposal for train ferries, largely because of the problem in conveying a full-length train in a ship, and overcoming the handicap of tidal rise and fall by constructing very costly enclosed docks. They concluded that a ferry was no real substitute for a tunnel – something that history was to support.

The promoters of the tunnel who gave evidence to the sub-committee included Sir Arthur Fell, Sir William Bull, Baron Emile d'Erlanger, (chairman of the Channel Tunnel Company) and the Southern Railway's chief engineer, George Ellson who, following in the steps of Brady and Tempest, was also the tunnel company's chief engineer.

The scheme prepared by the two Channel Tunnel companies and

agreed with the Southern Railway and the Nord Railway, was estimated to cost £30 million. Changes of traction from steam to electric would be involved at the entrances to the tunnel. There was also a problem of the difference between the British and French loading gauges, namely the clearances in both height and width needed to accommodate rolling stock, though the track gauge in Britain is for practical purposes identical with that on the Continent. Enlarging the Southern's loading gauge would have cost, at 1930 prices, about £10 million. The Southern was not able to accept such a commitment and eventually the Nord Railway, having consulted the other Continental railways concerned, agreed that special rolling stock conforming to the British loading gauge but suitable for Continental operation in such matters as couplings, brakes and buffing gear, would be built.

The whole policy of the Southern towards the tunnel was necessarily affected by the fact that it had much more extensive shipping interests than the Nord Railway. The general manager, Sir Herbert Walker, was inclined to give the project cautious support but he felt obliged to consult his board (then chaired by General Sir Everard Baring) and receive their instructions before giving evidence to the committee. Sir William Bull in fact complained to d'Erlanger about the Southern's lack of enthusiasm. D'Erlanger told the annual general meeting of the Channel Tunnel Company on 30 June 1936 that he had met Baring and Walker and that 'whilst still professing a friendly attitude towards the project they made it clear that in their opinion the interests of the Southern Railway must suffer from construction of the tunnel'.

It was unfortunate that a director of the Southern Railway, Lord Ebbisham, who was also a member of the economic advisory council committee, was sufficiently hostile to the tunnel to write a memorandum of dissent to the economic advisory council committee's report when it was published. Walker's assistant, Gilbert Szlumper, who was to succeed him in 1937, was also a fierce anti-tunneller – perhaps due to his long association with the docks and shipping side of the Southern Railway. He gave a paper on Cross-Channel transit attacking the tunnel, which provoked Sir William Bull into replying in a letter to *The Times*. Walker's own attitude has been described by J. L. Harrington, then one of his aides, and later general manager and deputy chairman of British Rail's Shipping & International Services: 'As a sound international railwayman Walker believed in the tunnel, but he had to take account of its effect on the Southern. He knew that further extensions of electrification depended on (a) the ability of the Southern to raise extra capital for these projects, which in turn depended on (b) the ability of the Southern to pay a dividend on the ordinary stocks, which again depended on (c) the profits from the shipping services and the port enterprises.'

Walker's evidence to the committee stressed the problem of the two loading gauges. It also pointed out the loss of receipts from the shipping services, and the need to write off capital expenditure already incurred on Dover Marine station and Folkestone Harbour. Finally, while the Southern would support the scheme if the government decided that the tunnel should be built, it would expect to receive compensation for the loss of revenue. The latter claim is interesting in view of the large increase in Continental rail traffic which all later studies of the tunnel have anticipated.

The committee took much technical engineering evidence covering excavation, spoil removal, ventilation and drainage, from consulting engineers, including the results of French studies. On the economic cases, there were different traffic estimates submitted by the Channel Tunnel Company, the Southern Railway and the Nord Railway. The Channel Tunnel Company was the most optimistic in regard to both traffic stimulation and diversion from the sea routes to the tunnel. The Nord Railway was rather more conservative, and the Southern Railway, with its very large shipping interests, was (not unnaturally) even more conservative. The committee compromised; its estimate was less than the tunnel company's figure, but more than those of the two railways.

The wording of the report was 'on the whole' encouraging. 'On the whole, after weighing all the evidence, we are of the opinion that the advantages that would accrue to this country from a Channel Tunnel would be likely to increase with the passage of the years. . . . We conclude therefore on balance that, if, as we believe, a Channel Tunnel could be built, maintained and operated by private enterprise at a cost which would permit of the traffic through it being conveyed at rates no higher than those at present in force on the short cross-Channel route, its construction, by creating new traffic and thus increasing trade, would be of economic advantage to this country'.

The committee was cautious about the financial prospects; it thought that the return on the capital invested, though probably 'meagre' in the early years, should nevertheless 'increase steadily'. The committee was equally cautious in regard to construction costs and it pointed out that questions of defence and international law would have to be settled – glimpses of the obvious. Its report was signed on 28 February 1930.

The Prime Minister sought an escape from the need to take action on the sub-committee's report, transmitted via the Royal Commission, by consulting the Imperial Defence Committee at a meeting on 29 May

1930. The military simply re-affirmed their objection to a tunnel and Ramsay Macdonald on 5 June told the House of Commons that he saw no reason to change the position of previous governments. A white paper (*The Channel Tunnel: Statement of Policy* – a misnomer since it only amounted to an apologia for having *no* policy for the tunnel) – was produced to justify inactivity by raising feeble objections – uncertainty about the final cost, possibility that the government would be committed to seeing the tunnel built if it consented to licence the construction, stimulating British tourism abroad more than tourism in this country from abroad, cost of extra defence measures, and relief of unemployment being confined to the period of construction. It ended by suggesting that the economic advantages of the tunnel did not outweigh the objections.

The supporters of the tunnel, backed by widespread comment critical of the white paper, refused to accept defeat. They forced a debate in the House of Commons on 30 June 1930. It lasted only three hours and before it started the Prime Minister, Ramsay Macdonald, announced a free vote but ominously declared that as the government had already decided against the Channel Tunnel, it would not feel bound by the result of the voting even if that should be in favour. Not a very convincing display of democracy in action!

The debate was opened by Ernest Thurtle (Labour, Shoreditch) who moved that 'since the Channel Tunnel can be constructed by private enterprise without any financial assistance from the state, and since the Channel Tunnel committee has reported its construction to be of definite economic advantage to this country, and, in view of the fact that such a tunnel, in addition to providing immediate employment, would be of great advantage to British trade and industry in future years by providing better transport between this country and the Continent, every facility should be given for the project to be undertaken at the earliest possible opportunity.'

Thurtle said that widespread popular support for the tunnel had been met by an attitude of permanent officialdom steadily against it; he forecast a revolt against 'that kind of dictatorship by the bureaucracy'. He emphasised the economic arguments and (surprisingly for a Socialist) agreed that the Southern Railway would be 'amply' compensated; he derided the military objections. Sir Basil Peto (Conservative, Barnstaple) seconded the motion. Ramsay Macdonald then made a lengthy speech which displayed not merely a lack of enthusiasm but a reiteration of every objection that civil servants and service chiefs could think up. He asked the House to support the government's decision to adhere to the policy of do-nothing which had persisted for nearly 50 years.

The subsequent speeches were in general neither constructive nor detailed; for instance Colonel Wilfred Ashley (Conservative, New Forest and Christchurch) said that 'the only argument he had ever heard in favour of the tunnel was that no one wanted to be seasick'. Major Hills (Conservative, Ripon) opposed the tunnel; but it was supported by Fenner Brockway (Labour, Leyton East) Noel Baker (Labour, Coventy), and Captain H. Balfour (Conservative, Isle of Thanet).

Herbert Morrison, the Minister of Transport, wound up for the government. He argued that if the tunnel were to be built, it should be done by the state, or the two states involved, rather than by private enterprise. He thought the economic case was weak and suggested that the only real advantage would be avoidance of seasickness – a surprisingly negative contribution by the energetic creator of the London Passenger Transport Board.

The result of the voting was

For the resolution	172
Against	179
	—
Majority against	7

On the same day, at the Channel Tunnel Company Annual General Meeting, Baron Emile d'Erlanger denied that the tunnel was dead – that, he said, was not his 'swan song'.

The two railway companies turned almost immediately to consider a train ferry service. Both Dover and Richborough were inspected as possible sites for the English terminal. Richborough–Calais was a route seriously considered, and Great Eastern Train Ferries, which was already operating freight train ferry services between Harwich and Zeebrugge, was interested in a Harwich–Calais service. But by the end of 1932 the Southern Railway was able to announce that it had agreed with the Nord Railway on the inauguration of a train ferry between Dover and Dunkirk.

The possibility of a through London–Paris sleeping car service in addition to the freight wagon business was mentioned. The train ferry dock at Dover was estimated to cost £545,000. Three train ferry ships were ordered from Swan Hunter and Wigham Richardson Ltd, each 360ft long with a speed of 16½ knots. They were named after Thames ferries – *Twickenham Ferry, Hampton Ferry* and *Shepperton Ferry*. After delivery, *Twickenham Ferry* was transferred to the French registry and sailed under the French flag.

In 1936 the Wagons-Lits Company provided 18 sleeping cars built to the British loading gauge and the Night Ferry was inaugurated between London and Paris in October of that year. The Night Ferry certainly had its glamorous side; it was intriguing to see the blue sleeping cars of the *Compagnie Internationale des Wagons-Lits et des Grands Express Européens* in Victoria station next to the Southern's prosaic electric trains, and the Wagons-Lits conductor in his chocolate-brown uniform standing on the platform to check his passenger list – which included many notabilities over the years, especially the Duke of Windsor who appreciated the discreet shielding from publicity that the Wagons-Lits staff were trained to provide for distinguished passengers.

But the service was far from ideal. The transfer from land to ship was protracted and involved much shunting in the night hours. Then the sleeping cars had to be secured with heavy chains against the motion of the ship: midnight clanking noises disturbed the passengers. And a rough night in the Channel was just as nauseating in a sleeping car as in a ship's cabin.

The Night Ferry train was notorious for its unpunctual arrivals, due either to weather conditions, problems in passing the locks at Dunkirk, or finding pathways into London during the morning peak commuter period.

The freight ferry wagon business was more financially important to the railways than the Night Ferry. The total fleet of wagons was large. But turn-round times were excessive and shipping space was by no means always available as wagons arrived at the ports.

Making do with second best has often been a British characteristic, and the train ferries were accepted as substitutes for a tunnel that would probably never be built. The tunnel receded from the public consciousness during the years when Britain slowly emerged from the worst of the depression and then belatedly in the late 1930s began to re-arm against the growing threat of Nazi Germany.

Sir William Bull died in January 1931 after many years of fighting for the tunnel ever since he had joined Fell's committee in 1913; he had been Fell's chief lieutenant until he succeeded as chairman in 1922. Sir Arthur Fell died a few years later, in 1935. His granddaughter has written of him: 'He was a great Francophile and much loved by all the French, which was natural because although educated at King's College in the Strand and St John's College, Oxford, his mother, when not hobnobbing with Ellen Terry in the Earl's Court Road, lived in a flat in the Rue St Honoré in the winter, Biarritz at Easter, and Avranches in the summer'. His obituary by his opposite number in France, M. Bertin, in the *Journal des Debats* on 14 January 1935 ended: 'Hommage à Sir Arthur Fell!'

Baron Emile d'Erlanger retired from the chairmanship of the Channel Tunnel Company in 1937, and was followed by Sir Herbert Walker on the latter's retirement from the general managership of the Southern Railway. In this capacity Walker was able to devote more time and sympathetic attention to the tunnel than had been possible when he was pre-occupied with the position of the Southern Railway. Sir Herbert remained with the company until his death in 1949; his co-directors at first were Lord Radnor, a director of the Southern and a very prominent figure in Folkestone, and later Leo d'Erlanger. Eventually John Elliot, deputy general manager (and briefly in 1947 acting general manager of the Southern) replaced Lord Radnor.

The second world war dramatically changed the whole situation in which the British military establishment had been able to argue that the tunnel would be a threat to national security. Air warfare made this obsolete. On the contrary, once again, as in 1914–18 the desirability of a tunnel was urged by the French. The Chamber of Deputies in November 1939 during the period of the 'phoney war', when the British Expeditionary Force was once again helping to man French defences against Germany, passed a resolution demanding construction of a tunnel. The British Prime Minister, Neville Chamberlain, was the son of the Joseph Chamberlain who as President of the Board of Trade had ordered Watkin to stop work in 1882. He upheld the family tradition in a parliamentary answer rejecting the tunnel.

It was strongly argued by many in Britain that the tunnel would have been a great asset in speeding up the evacuation of the BEF after Dunkirk. On the other hand, the former fear of the French seizing the British tunnel entrance and using it for an invading force was now replaced by a suggestion that German paratroopers might have been able to do the same, had the tunnel been in existence. The War Office did not entirely exclude the possibility that the Germans might seek to invade Britain by surreptitiously building a tunnel at great speed after the fall of France. This rather far-fetched notion was taken seriously enough for an eminent consulting engineer, Sir William Halcrow, to be asked whether such work could be carried out without detection. His conclusion was that the arrangements for spoil disposal would have to be so substantial that air reconnaissance would soon detect them. However, air reconnaissance was in fact continued on the site of the closed French tunnel workings at Sangatte and there was a flurry of interest in early 1944 at some signs of unusual activity there; it later appeared that this was concerned with launching sites for V2 rockets.

Preparations for the Allied invasion of France did not exclude a further look at the tunnel, but it was quickly decided that the time-scale for construction ruled it out. Meanwhile the Germans used the old shaft at Sangatte as a rubbish dump until they finally decided to cap it with concrete. They then even tidied it up with stone steps and paths, to be in harmony with their adjacent military cemetery.

The Channel Tunnel Company's offices in the Southern Railway headquarters at London Bridge station were destroyed in an air raid and many of its historic records were lost, including the register of shareholders. The Germans certainly had done their best to kill off the tunnel; its revival in the 1950s was the result of a surprising combination of events.

6 The Project Reborn

Soon after the war ended in 1945, interest in the tunnel revived in a number of quarters. As early as 17 March 1947 Christopher Shawcross, a young Labour MP, took the lead in arranging a meeting at the House of Commons which agreed to revive the Channel Tunnel Parliamentary Group, with a remit to look at the project in the light of up-to-date engineering practice and revised traffic estimates, and to report. For this purpose the group set up a drafting committee under the joint chairmanship of Shawcross and Captain Malcolm Bullock, MP, with Lt Commander Christopher Powell, RN, as secretary.

Sir Herbert Walker, who had retired as general manager of the Southern Railway a decade earlier was still chairman of the Channel Tunnel Company which was in considerable disarray owing to the loss of its office and many of its records in the blitz. His interest in the tunnel may have been strengthened since he had become a director of The United Steel Companies Ltd. If the tunnel were built it would probably be lined with cast-iron segments similar to those used for the London Underground. United Steel had an iron foundry at Wellingborough which would be interested in such work.

Sir Herbert was in contact with the revived parliamentary group until his death in 1949 and was responsible for putting George Ellson, who had been chief engineer of both the Southern Railway and the Channel Tunnel Company, in touch with the group on technical questions. Ellson advised that the project of 1930 was basically sound but suggested that if rolling stock to 'train ferry' instead of Continental loading gauge were accepted by the French railways, the tunnel's diameter could be reduced from 18ft 8in to 17ft. He also suggested using reinforced concrete tunnel segments instead of cast iron.

Ellson was once about to leave for a tunnel board meeting when his new works assistant burst into his room with the news that a major fault had been discovered in the Lower Chalk strata of the sea bed. Ellson left, considering how to break the news of this unhappy discovery, but meanwhile his assistant, checking and re-checking the figures, discovered that an error had occurred, in the conversion of metres into feet. Just in time, a telephone message reached Ellson to the effect that the Channel bed was after all intact!

The Channel Tunnel parliamentary group's committee produced its report in July 1947. It updated the 1930 estimates to take account of a rise in estimated construction costs, from £30 millions to £65½ millions. The traffic estimates were also amended to reflect the growth of air traffic. Even so, the tunnel was estimated to be profitable and the group ridiculed the hoary military objections in the new age of rockets and atom bombs. If the military arguments had really been valid, the international tunnels through the Alps would surely not have been built. Perhaps owing to the political circumstances of the period, the group argued that the tunnel should not be privately financed but be publicly owned, as a nationalised industry. The Labour government in Britain was nationalising virtually all public transport, as from 1 January 1948; the French Railways had been nationalised since 1937.

The military objections to the tunnel now appeared increasingly irrelevant and outdated, though, suprisingly, Lord Montgomery was to describe the tunnel as a 'wild-cat' scheme, seeming to echo Lord Wolseley when he rhetorically demanded in a speech to the Navy League in 1957 'why give up one of our greatest assets – our island home?' But, in reply to a question from E. L. Mallalieu, MP, then joint chairman of the Channel Tunnel Parliamentary Group, the Minister of Defence, Harold Macmillan, stated in 1955 that, whatever retired generals might say, the Ministry of Defence now considered that there were 'scarcely any' adverse military implications in the tunnel. Indeed, for a time the post-war military establishment in Western Europe – Supreme Headquarters Allied Powers Europe (SHAPE) – had considered that Western defences against the Eastern bloc would be greatly strengthened by a Channel Tunnel. SHAPE studied with interest a tunnel scheme devised by a French engineer, André Basdevant, that would provide a four-line motorway with a central separation, and above it a two-track railway. This huge tunnel was to be in the form of an ellipse, not a circular bore, and to be progressively enlarged during construction, from a pilot tunnel. The route would be rather to the south-west of the usual one, passing under the Varne bank. The tunnel motorway would have joined the Folkestone–London road near the village of Newington and the railway (surprisingly) would have joined the Elham Valley line from Folkestone to Canterbury near Lyminge, so that the old LCDR route to London would be used rather than the old SER line via Ashford for the international trains.

The attractiveness of a tunnel with such a large throughput capacity was obvious to the military at SHAPE but the engineering problems in Basdevant's scheme were formidable. First of all, ventilation of a tunnel motorway of this length, with the forecast density of traffic, posed technical questions still unsolved. Then, Basdevant's route took the tunnel out of the Lower Chalk and into the (probably) water-bearing greensand on the English side – in fact it was not based on geological information comparable with that of the normal route planned since 1930. Its rail connection on the English side was with a single-line branch which the railway management was in process of closing. The cost would be enormously greater than that of the twin-bore railway tunnel.

In March 1948 the British and French tunnel parliamentary groups had a meeting at which prominent engineers, including George Ellson, now chief engineer of the Southern Region of the newly nationalised British Railways, and M. Gonon, chief engineer of the Nord Region of French Railways, took part in examining Basdevant's scheme. The result was a preference for a rail tunnel with twin bores, though Basdevant's design was considered to be worth further examination. Despite leaving the door ajar in this way, the Basdevant tunnel gradually lapsed into obscurity; its main importance is that it was the first scheme to take account of the need to provide adequately for cross-Channel road transport. The drive-through design or tunnel motorway has some obvious attractions as compared with relying upon rail technology, and indeed it was still being actively considered as late as 1985. It always seems to founder on three huge question-marks: geological problems created by the large diameter; ventilation; and cost.

The early 1950s saw a number of important developments. The Shawcross group's report had stimulated interest in France where a parliamentary study group was formed with M. Guy d'Arvisenet as secretary. Then, the British Channel Tunnel Company acquired a new chairman in Leo d'Erlanger, a member of the great international banking family and grandson of Baron Emile who had held the fort in earlier days. D'Erlanger was a charming and remarkable man, full of complexities. When he died in October 1978, his obituary in *The Times* described him as 'a man born to wealth who lived off toast and furnished his bedroom like a cell . . . a devout Catholic given to the quaintest superstitious practices'. He had, after a brief glance at a lady, immediately followed her round the world and married her. He was to take his place in the long succession of notable fighters for the tunnel, ranking with Thomé de Gamond, William Low, Sir Edward Watkin, Sir Arthur Fell and Sir William Bull.

D'Erlanger was accustomed to say that he had reserved a seat in the first train to pass through the tunnel; but if it should not be open before his death, he had left instructions that his coffin was to be dug up and placed in the leading guard's van of the train.

At the outset he had taken over the chairmanship without enthusiasm, largely as an act of family piety and respect for his grandfather. He later recalled that the Channel Tunnel Company in those first post-war years was badly in the doldrums. Its annual general meetings, held in the Charing Cross Hotel, drew only a tiny handful of shareholders and d'Erlanger remembered that originally he had looked forward to them with dread, because he had nothing to say and the project was moribund. There was even a problem in getting a quorum for the meetings.

Gradually however d'Erlanger began to develop an interest in the tunnel and this was stimulated in 1956 by a meeting with a French friend, Paul Leroy-Beaulieu, financial counsellor at the embassy in London since 1953, who became inspecteur-géneral des finances in 1956. They recalled that their grandfathers had been simultaneously chairmen of the English and French Channel Tunnel companies.Leroy-Beaulieu was in fact the grandson of Michel Chevalier who had been the moving spirit in the French Channel Tunnel Company (the 'Société Concessionaire') of 1875. Their discussion included the possibility of interesting the Suez Canal Company in the tunnel, since that company's concession was due to expire in 1968 and might not be renewed.

A chance conversation about this time also had a significant effect. Two French sisters, living in the USA, and married to prominent businessmen, came to Europe in 1956 for a holiday and experienced the misery of a really bad Channel crossing. They asked their husbands why no one had built a Channel Tunnel, and the question was taken up seriously by their spouses, Frank P. Davidson and Comte Arnaud de Vitry d'Avaucourt. These two were influential and active in USA financial and legal circles. Frank Davidson was, though comparatively young (born in 1918) a successful lawyer at the New York bar; de Vitry d'Avaucourt, also young (born 1926) was a senior executive of the Socony Mobil Oil Company. After getting in touch with the English and French Channel Tunnel compaines, they sent a lawyer, Professor Cyril Means, Jr, to London and Paris in February 1957, where contact was made not merely with the tunnel companies but also the Suez Canal Company.

The Suez company had already given some tentative thought to investing in the Channel Tunnel. Georges-Picot, director-general of the company, had been trying unsuccessfully to negotiate with Colonel Nasser for compensation after the canal was nationalised by the Egyptian leader; but even without that, it possessed large financial reserves looking for profitable investment, and the tunnel seemed a good field for the exercise of skills akin to those needed for the management of an international waterway.

Cyril Means's visit to Europe was sufficiently fruitful to arouse American interest. In March 1957 a corporation was formed in New York with Comte Arnaud de Vitry d'Avaucourt as President, entitled Technical Studies Inc, its functions defined as being to finance technical investigations and promote the construction of a Channel Tunnel.

At the 76th annual general meeting of the British Channel Tunnel Company Leo d'Erlanger was able to announce that the British and French companies were henceforth to collaborate with the Suez Canal Company and also Technical Studies Inc. Intensive discussions followed between the parties and on 6 July 1957 the four participants were able to report that they had combined to set up a Channel Tunnel Study Group in which Suez had 30%, the British Channel Tunnel Company 30%, the French Group 30% and Technical Studies Inc 10%.

The members of the Study Group, nominated by the four participating bodies, reflected a wide range of interests. Georges-Picot represented the Suez Canal Company and Leo d'Erlanger the British Channel Tunnel Company. However, the nationalised British Transport Commission had inherited from the Southern Railway a block of 120,000 Channel Tunnel Company shares, and the chairman of the Commission, General Sir Brian Robertson, was a strong supporter of the tunnel. Accordingly, a member of the Commission, A. B. B. Valentine, was appointed to the Study Group with a watching brief. Valentine's background had been in urban transport. He had been a manager with the Underground railways of London and with the London Passenger Transport Board. After 1948 he had been a member of the London Transport Executive and since 1953 a member of the British Transport Commission.

On the French side the team was a strong one, reflecting the various interests. The French company's shares were owned in the proportion of 25% by de Rothschild frères, 50% by the French Railways (SNCF) and 25% by minority holders. Louis Armand, president of the French Railways and at the time also president of the UIC (International Union of Railways) was included, as was a director of the French Channel Tunnel Company, Jacques Getten, who was also a partner in de Rothschild frères, reflecting that bank's long-standing interest in both the tunnel and (formerly) the Nord Railway. For the first time the importance of providing for the interest of road traffic was acknowledged by nominating Baron Charles de Wouters d'Oplinter, a

Belgian banker and industrialist who was also President of the Fédération Routière International. Technical Studies Inc was represented by Comte Arnaud de Vitry d'Avaucourt, George W. Ball, a lawyer, and Thomas S. Lamont, a banker and partner in the house of Morgan.

Later the British and French companies and the Suez Canal Company each reduced their participation to 25%, and the share of Technical Studies Inc was increased from 10% to 25%.

With this formidable team assembled, the way was clear for work to start on a project that came within an ace of proceeding to completion in 1974. But 18 years of struggle lay ahead before the point was reached – years in which stop-go seemed to dominate almost until the final stage was reached.

7 A Truly Viable Scheme

The first-fruits of Cyril Means's visit to England and France in February 1957 were a revival of activity by the British Channel Tunnel parliamentary committee and a request from the committee to Brian Colquhoun & Partners, a firm which had acted as consulting engineers in several tunnel projects including the Mersey Tunnel and the Maas Tunnel in Rotterdam, for a preliminary report on the present state of the Channel Tunnel. This report was quickly made, in April 1957. It outlined a plan of work covering fuller study of the geological structure of the sea-bed, but it agreed that the Lower Chalk appeared an ideal tunnelling medium and that Low, Hawkshaw and Fox had all been right in their choice of route as against Basdevant and de Gamond. Colquhoun suggested testing the sea-bed by the 'sonoprobe' method and by the sinking of further boreholes. He outlined various possibilities for the actual tunnel design, which should be dictated eventually by the results of exhaustive analyses of present-day cross-Channel traffic and forecasts of its development. He left open the question of a single tunnel (following Hawkshaw) or twin tunnels (following Low) and also the possibility of integrating roadways with the railways in the tunnel.

The Colquhoun report was made available to Technical Studies Inc and the other members of the Channel Tunnel Study Group. The Group set up a supervisory commission to oversee the planning, with highly distinguished British and French co-chairmen – on the British side, Sir Ivone Kirkpatrick, GCB, a retired diplomat whose wide experience, including posts as UK High Commissioner in Germany and Permanent Under-Secretary at the Foreign Office, made him an excellent counterpart to his French colleague, H. E. René Massigli, a distinguished French engineer and a former French Ambassador to Britain. Under the commission, the work started in the hands of René Malcor, an experienced French civil and military engineer who had been director of public works in the military government of Germany between 1945 and 1950. He was also an expert on soil mechanics and traffic engineering. He undertook to organise the geological survey, the traffic survey and the engineering design.

The major product of the work commissioned by the Study Group was a massive report of March 1960 which can still be regarded as the fundamental tunnel plan since it was the real basis of the scheme accepted in principle between 1966 and 1974 and suddenly abandoned in 1975; and even of its successor, the scheme selected in 1985. In fact, between 1957 and 1960 the whole ground work was laid; it has subsequently been modified more in detail than in fundamentals. The hesitations, doubts, stoppages and restarts since 1960 reflect nothing more than the lack of political will, a shortcoming more apparent on the British than the French side, though there have been some changes of heart in Paris from time to time which have caused temporary delays.

The Study Group determined to obtain the best professional advice on the one hand, and on the other to consider not merely a bored tunnel but also the possible alternative forms of fixed link that had been proposed from time to time. The principal traffic and economic consultants were the Economist Intelligence Unit, the French Société d'Etudes Techniques et Economiques (SETEC), and the USA firm of DeLeuw, Cather & Co. H. J. M. (Sir Harold) Harding was appointed British consultant in engineering aspects; he was vice-president of the Institution of Civil Engineers in 1959 and in addition to his great practical and academic attainments in engineering he was a noted specialist in soil mechanics. Major studies were commissioned from the Société Générale d'Exploitations Industrielles (SOGEI), from Sir William Halcrow & Partners, Livesey & Henderson and Rendel Palmer & Tritton, all leading engineering consultancy firms. Two geophysical consultants, Professor J. M. Bruckshaw of the Imperial College of Science & Technology, and Professor Jean Goguel of the Sorbonne, were also appointed.

This formidable array of experts had a great mass of data accumulated over the years to sift through. One of the most unexpected finds came in a closed Paris suburban railway station, used as a store, where some thousands of samples of the Channel sea-floor dating from the studies of 1875 and 1876, together with original specimens recovered in 1855 by Thomé de Gamond from his naked diving exploits, turned up. Generally, these specimens – all neatly labelled – agreed with the results of later and more sophisticated studies.

The Study Group took care to forestall any criticism that they had committed themselves to any particular type of fixed link before commencing their investigations. They studied, in addition to a bored tunnel, an immersed tube, a bridge, and a combined bridge and

The substitute for the Tunnel: the northbound Night Ferry train with Paris–London sleeping cars near Bickley Junction. (*British Rail*)

One of the three Dover–Dunkirk train ferry ships, *Shepperton Ferry*, leaving Dover. (*British Rail*)

(*left*) The great polymath, Thomé de Gamond, apparently brooding over his Tunnel concept. (*Author's collection*); (*right*) One of the greatest English Victorian engineers, Sir John Hawkshaw, an enthusiastic Tunnel designer. (*Author's collection*)

Sir Edward Watkin of the South Eastern Railway (wearing fur hat) in an unusual setting: the special train for directors to inspect the site of the great landslip in the Warren between Folkestone and Dover in 1877. (*Kent County Library*)

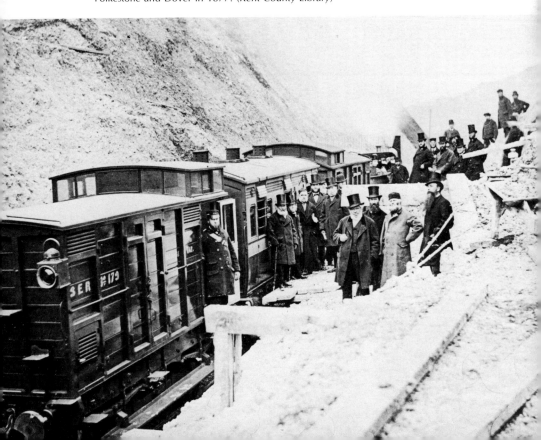

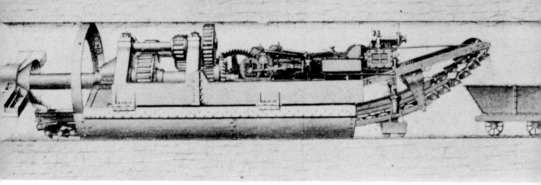

The Beaumont tunnelling machine of 1882. (*Eurotunnel*)

The 'Hawkshaw' workings at St Margaret's Bay. (*Author's collection*)

The Shakespeare Cliff Colliery, later established near the site of Sir Edward Watkin's Channel Tunnel workings. (*British Rail*)

An entrance to the 'Watkin' Channel Tunnel kept on a 'care and maintenance' basis for almost 90 years. (*British Rail*)

Inside the 'Watkin' Tunnel, the chalk standing firm though unlined after 90 years. (*British Rail*)

(*left*) For many years the chief champion of the Tunnel in Parliament, Sir Arthur Fell, MP. (*Courtesy Mrs Lorna Corin*); (*right*) Perhaps the best-known of all Tunnel champions, the Rt Hon Sir William Bull, PC, MP. (*Courtesy Mr Anthony Bull*)

(*left*) Converted to enthusiasm for the Tunnel: Sir Herbert Walker, the great general manager of the Southern Railway (*David Lipson*); (*right*) A great Tunnel advocate, that remarkable and distinguished banker, Leo d'Erlanger. (*Courtesy Mr R. F. d'Erlanger*)

Cross-section of André Basdevant's proposed enormous road and rail tunnel. (*Author's collection*)

A cut-away model of the Basdevant tunnel scheme. (*Author's collection*)

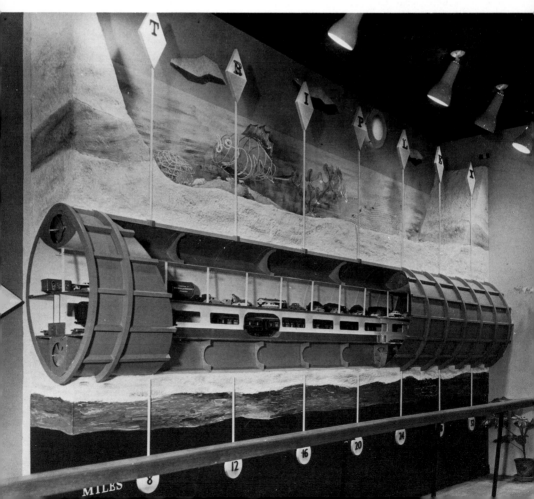

(*above*) New government enthusiasm in the 1970s: the Minister (John Peyton, second from right) visiting the old Channel Tunnel workings. (*British Rail*); (*below*) The entrances to the two access tunnels, on the Shakespeare Cliff site, after abandonment in 1975. (*Eurotunnel*) (*overleaf*) The main access tunnel in 1974. (*British Rail*); (*inset*) The French entrance to the 1974 workings – 'La Descenderie'. (*Eurotunnel*)

Model of interior of single-deck wagon for coaches and lorries. (*Eurotunnel*)

Model showing shuttle ferry trains entering and leaving the Tunnel. (*Eurotunnel*)

Approaches to Victoria: the congested site ruled it out as the London terminal for Channel Tunnel services. (*British Rail*)

Waterloo approaches, showing where the widening for the international station will be undertaken alongside the existing viaduct. (*British Rail*)

Site of the former spur direct from Waterloo lines to the West London Line, closed in 1916, which will be reinstated for Channel Tunnel trains to and from North Pole servicing centre. (*British Rail*)

Waterloo's 'Victory Arch' – which should also become the 'Gateway to the Continent', a status held by Victoria for over a century. (*Author's collection*)

Waterloo International Terminal – an artist's impression. (*British Rail*)

tunnel. They came to the conclusion that the bored tunnel was, both from an engineering and an economic standpoint, the most effective fixed link but they were prepared to continue study of an immersed tube. They built upon the experience and studies of the past, and proposed two single-line rail tunnels with a diameter of 21ft 4in, with between them a service tunnel 7ft 6in in diameter. The service tunnel would be excavated in advance of the running line tunnels and would have frequent cross-passages to assist access for maintenance, for ventilation and also provide a safety feature as a means of evacuation in an emergency. The total length would be 32 miles or 51 kilometres, of which 23 miles (37 kilometres) would be under the sea.

The vertical profile would be in the form of the elongated 'W' first suggested by Sartiaux and Fox, with drainage effected by pumping from the low points of the 'W' near the coasts. The gradients from the terminals to the low points of the 'W' would be 1% (1 in 100), and from the low points to the central high point no more than 0.1 per cent or 1 in 1000.

The terminals would be at Sangatte near Calais and Westenhanger in Kent. Each terminal would comprise not merely a frontier station for through trains between England and France, but also a road-rail interface, since it was proposed to carry cars, lorries and coaches on ferry trains running on standard gauge rails, at frequent intervals according to the demand and the time of day. The ferry train wagons would be either double-deck (for cars) or single-deck (for coaches and lorries) and consequently would be wider and higher than any in service either in Britain or the Continent: the tunnel's diameter would be dictated by their dimensions.

The ferry trains would run in a continuous circuit without reversal, using loops at each terminal. The estimated time between entering one terminal and leaving the other was put at 65 minutes, of which 45 minutes would be the journey proper. A five-minute interval between ferry trains would allow a theoretical through-put of 3,600 vehicles an hour in each direction.

The Study Group set out the following estimates of traffic.

	1965	1980
Accompanied vehicles	675,000	1,130,000
Passengers	3,200,000	4,850,000
Goods (tons)	1,753,000	2,609,000

The costs were estimated as follows:-

	£ million
Tunnel proper	80.0
Terminal installations (excluding railway equipment)	14.0
Fixed railway equipment	9.8
French highway construction	2.2
Ferry wagons	6.0
	112.0

The plan of finance was to raise £160 millions by a share issue of £32 millions and loans amounting to £128 millions. The loans would be long-term (25 years) and medium-term (9–15 years) in the form of bonds and bank credits, the bonds carrying conversion rights into shares. The share capital would be raised 40 per cent in France, 40 per cent in Britain, and 20 per cent in other centres. Bank credits were expected to be available in London and New York.

The Group proposed that the project should be regarded as a commercially viable one, to be financed and managed by an Anglo-French Tunnel Operating Company, with a concession from the two governments for a period of 99 years. It also asked that, in order to assist raising the necessary finance in the two countries, the British and French governments should guarantee the interest on the bonds. The prospective rate of return on the equity capital was put at 10% in the first year of operation, thereafter increasing at 3% per annum. Of the total of 10%, the Group suggested that 8% might be paid as dividends and 2% put to reserve.

The Group's proposals were set out in detail in two massive volumes, one containing the engineering and technical appraisal, the other the economic and financial case. It might have been expected that they would be well regarded by the two governments, and that the issue was sufficiently clear-cut for a decision, one way or the other. But it always seems to happen throughout the tunnel's history that consideration of a project is complicated by the emergence of rivals claiming to deserve equal study. The water, in fact, quickly becomes muddied.

That occurred soon after publication of the Study Group's proposals, and it must be said that the rival scheme suddenly put forward did not represent the lunatic fringe of fixed-link proposals; it had quite powerful backing. Once again it incorporated a bridge, even more substantial than the one examined by the Study Group and rejected by them, a multi-purpose steel viaduct carrying two railway lines, five lanes of road traffic and (a nice touch) two bicycle tracks. It would be 21 miles long, crossing the Channel directly from Dover to Calais, and would be 115ft wide and 230ft above high-water level. The spans would be 660ft long, resting on 164 concrete piers 65ft in diameter.

The principal sponsor of the bridge was of course the steel industry. Both in Britain and France that industry has been more interested in bridge projects even though a tunnel would probably require cast-iron segments which many major steel firms would be able to produce at some of their associated foundries. Ever since George Ellson's report of 1948, however, the alternative possibility of using reinforced concrete segments had become steadily more likely to be chosen and the steel industry has accordingly thrown its weight behind a bridge rather than a tunnel.

The 1960 proposal, apart from its support from industry, gained authority from being put forward by a company presided over by M. Jules Moch, a former Minister of the Interior in France, and supported by Lord Gladwyn in Britain. The bridge was in the direct line of succession from the scheme of Schneider and Hersent in 1889, which had been supported by the French Ministry of Public Works but ruled out by the Ministry of Marine. Another ancestor was an older proposal of 1930, supported in Britain by Sir Murdoch Macdonald & Partners.

The Moch proposal had sufficient backing for the two governments to decide that it must be examined alongside the Study Group's tunnel scheme. They agreed on 17 November 1961 to set up a working group of French and British officials to report on the two projects – whether they were technically adequate and practicable; whether they had serious implications in the field of international law and defence; how far, if at all, they would be an improvement upon the existing means of crossing the Channel by sea and air; and how far they appeared financially viable as private ventures.

The two co-chairmen of the working group were D. R. (later Sir David) Serpell and J. Ravanel, both very senior civil servants. The report (*Proposals for a Fixed Channel Link,* Cmnd 2137, in Britain) appeared in September 1963. It is often referred to as AF (Anglo-French) 63. On technical problems, it concluded that the bored tunnel was based on the application of proved techniques and that the risk of encountering geological difficulties that might prove fatal was 'negligible', though £1 million should be spent on supplementary soundings and seismic studies. They felt unable to assess the merits of an immersed tube alternative.

On the bridge, the working group felt that the project was technically practicable, though substantial modifications to the design would be needed to protect the piers against the risk of collision by ships. (The age of monster tankers, it may be commented, had not yet arrived.) But the real objections to any bridge were the hazard to shipping in one of the busiest shipping channels in the world. International agreement would be needed, involving all the major maritime nations, before such obstruction to international waters could be allowed. Special measures to regulate shipping in bad weather would be needed and possibly compulsory pilotage, all of which would be strenuously resisted by shipping interests. In conclusion, the group agreed that the bridge 'would constitute a serious new danger to shipping' and was 'open to very grave objections'. None of these problems arose with any form of tunnel, apart from interference with submarine cables which an immersed tube would involve.

Rather surprisingly, the group did not deal with other problems connected with the bridge such as the dangers to road traffic created by high side winds and, especially, the frequent Channel fogs.

Defence considerations, the group reported, were not considered by the military authorities whom they had consulted in Britain and France to be decisive, though the possibility of attack or sabotage could never be entirely ruled out from any form of fixed link. (Terrorist bombs, and the hi-jacking of ships and aircrafts, were not yet features of the contemporary scene in 1963.)

On the economic aspects, the group assumed a six-year construction period for each project and deliberately used figures of costs which were higher than those submitted by the promoters, because higher estimates were 'much less likely to be exceeded in the event'. They put the cost of the tunnel at £143 million and of the bridge at £298.5 million, and in the light of their traffic forecasts, found 'nothing that would seem to justify the choice of the project with the higher capital cost'.

The group's estimates of traffic differed from those of both the tunnel and bridge promoters; they used four bases, a higher and a lower level, subdivided into 'very low', 'low', 'high' and 'very high'. They accepted the levels of tolls proposed by the tunnel Study Group (those of the bridge group did not differ significantly). They considered traffic diversion from the established means and also the generated traffic which a new facility would create. Applying discounted cash flow techniques they found that the rate of return from the tunnel varied between 14.4 per cent on the basis of the 'very high' traffic estimates to 7.4 per cent on the 'very low' estimate. For the bridge the results were significantly worse, ranging from 7.1 per cent to 2.0 per cent.

For the overall economic assessment – the effect upon the national economy, expressed in money terms – the two schemes were compared with the alternative of continuing to rely upon the established means of crossing. The tunnel showed a positive gain under all assumptions; the bridge a much smaller gain on the 'very high' assumption, and a net loss on all the other assumptions.

The group's conclusions were that the bridge was less satisfactory than continuing to rely upon existing cross-Channel services; the tunnel was preferable to the do-nothing situation. More work needed to be done on geological studies, the role of the railways, the status of the tunnel operating organisation, and the question of a government guarantee of interest upon the bonds to be issued to finance the construction.

Subject to these cautionary words, AF63 seemed to point the way to acceptance of the Study Group's tunnel project and in fact only six months later the British and French Minsters of Transport announced on 6 February 1964 the decision 'in principle' to go ahead with the construction of a rail Channel Tunnel, and to examine further the problems posed by the project. These problems were stated to be primarily of a juridical and financial nature, but the government also wanted some more work to be done on the geological and geophysical aspects.

Given the announcement of February 1964, work continued, both on the part of the Study Group and by civil servants in both countries, along the lines laid down in AF63. In July 1964 the geological survey was started and it was completed by October 1965; 69 marine boreholes and 19 land boreholes were sunk. The Anglo-French Surveillance Commission agreed that the results confirmed the suitability of the route already proposed for the bored tunnel, but were not sufficient to establish the practicability of an immersed tube.

The civil servants however came to the conclusion that while construction could be carried out by private enterprise, the operation of the tunnel must be entrusted to a public body, a Channel Tunnel operating authority. This may have been in part consequential upon the change in the British Government's political complexion following the general election of 1964. The civil servants may have considered that the public operating authority would be more acceptable to a Labour Government than the possibility, mooted in AF63 under a Conservative Government, of a Treasury guarantee of interest upon bonds to be issued by a private-enterprise undertaking.

However, and with the benefit of hindsight, the concept of the public operating authority can be seen as a mistake. Not only did it complicate all the financial and administrative procedures that had to be sorted out, but it made absolutely essential the promise of a government guarantee of the bond interest payable by the construction company, since otherwise the bondholders would have no collateral security whatever once the tunnel had been handed over to the operating authority. That incidentally was to be a factor in the decision to cancel the project in January 1975 – the political and Treasury objections to putting public money, even in the form of a contingent guarantee, into the tunnel.

Apart from this possible error of judgement, the civil servants' conclusions, set out in a joint report by British and French officials usually referred to as AF66, were entirely encouraging. The geological problems were considered to have been satisfactorily disposed of on the lines indicated in AF63; the construction costs were re-assessed, as were the traffic estimates, the hypothetical tolls and the overall economic appraisal. The latter was re-calculated on two bases: 'most favourable' and 'least favourable', the latter allowing for both an over-run in construction costs and a shortfall in traffic. Even the 'least favourable' assumption, applying discounted cash flow principles, showed the tunnel to have a satisfactory internal rate of return and an adequate 'net present value' as an investment.

AF66 was not published but was circulated to ministers and senior civil servants. It was responsible for two announcements being made in 1966. On 8 July the Prime Ministers of Britain and France (Harold Wilson and Georges Pompidou) announced that it had been definitely decided that the tunnel should be built, subject to the details of how the work should be carried out being agreed on each side. A further joint announcement was made on 8 October by the British Minister of Transport, Mrs Barbara Castle, and the French Minister of Equipment, giving 1975 as the target date for the tunnel's opening.

At long last, the official green light! Or so it seemed. Now however began the stately minuet that was to last some half-dozen years, and seemed to involve orders of 'take your partners and advance; retreat and change partners; advance again; pause and retreat; finally advance and honour your partners'. From 1966 to 1970 the tunnel's progress, despite massive efforts by all concerned, seemed to be a matter of lurching rather than marching forward, meeting occasional showers of stones aimed by an odd assortment of opponents – vested interests such as the Dover Harbour Board, simple xenophobes who disliked the French, trade unionists (not only in the National Union of Seamen) who were suspicious of all contact with the Continent, and a collection of environmentalist groups.

The planning that had started in 1955 now had to be extended into many questions of detail, if the target date was to be met; it was shared between government officials, the tunnel promoters and the railways. Suddenly, BR and the SNCF were required to give much more serious consideration to their role in the tunnel.

8 The Role of the Railways

Following the Wilson–Pompidou announcement, civil servants in Britain and France began to work on the chief essential preliminaries to a tunnel, the legislation and the treaty. The question of relationships with the railways also became important. In the 19th century the tunnel had been regarded as a railway enterprise; in the 1930s, though an independent tunnel undertaking was envisaged, the railways were expected to be the sole users. Now the needs of road transport had become at least as important – probably more important – than those of the railways. Politicians and civil servants alike were anxious to avoid opposition from the powerful road lobby. The Study Group had acknowledged this by recruiting Baron de Wouters d'Oplinter of the International Road Federation, a body that in some ways paralleled the International Union of Railways.

The position of the French Railways, SNCF, in relation to the French Government was in some respects stronger than that of British Railways vis-à-vis the British Government. French tradition gave a status to major public utilities, a prestige and a confidence in dealing with governments, that British Railways lacked. With the appearance of deficits from 1955 onwards, BR had been subject to both political and Treasury pressure to economise; this had culminated in the 1961 appointment of Dr Richard Beeching, and his programme of sweeping reductions in the rail network. The change from a Conservative to a Labour Government in 1964 had, although slowing down the Beeching cutbacks, not materially improved the prospects for substantial government support of railway development. This was of course also the period when the fruits of the 1955 railway Modernisation Plan, particularly as regards electrification, were beginning to be realised and for the time being the railways were expected to live off their fat.

It was therefore made plain by civil servants in talks with BR officers that the railways' role in the tunnel would be clearly defined and circumscribed. Their technical expertise would be utilised – after all, no one in the world had so much experience of long tunnels as the railways. And electric rail traction had been decided upon in preference to a drive-through road tunnel on grounds of technical superiority, especially as regards ventilation, and cost. The railways would be the principal customer of the tunnel for use by their own trains, subject to the payment of the appropriate tolls, and probably would be also responsible for maintenance of the rail track and signalling, on a contract basis. Whether the operating authority should itself be a railway undertaking however was a difficult issue. It had early on been decided that the authority should own the ferry wagons; whether it should also own the locomotives, provide the train crews and be in charge of the signalling and control system of the tunnel was a problem left over for the time being.

An argument in favour of the authority itself carrying out the ferry train operating was that, as the prospective employer of the staff, it would be able to recruit on the basis of new contracts of service, and these might include a no-strike clause, all disputes being referred to a form of arbitration that would be binding on both sides. The road transport lobby had raised the bogey of either the National Union of Railwaymen, the Associated Society of Locomotive Engineers and Firemen, or any of the French railway unions being able to paralyse road vehicle traffic between England and the Continent. This of course ignored the fact that residual shipping services would almost certainly continue after the tunnel opened and, away from the tunnel's 'traffic shadow', such routes as Southampton–Le Havre and Harwich–Zeebrugge would be little affected by the tunnel. It also ignored the fact that shipping services themselves had been, and no doubt would continue to be, affected from time to time by industrial action on the part of the National Union of Seamen, the dock workers in the Transport & General Workers Union, and British customs and immigration officials or their Continental counterparts. If these facts were taken into consideration, the construction of the tunnel would be seen as reducing the likelihood of a breakdown in international connections, as there would be a wider choice of routes than had existed before.

From 1966 onward contacts between the Ministry of Transport and BR, and between BR and the SNCF, became closer and much more frequent. When Sir Brian Robertson had arranged for A. B. B. Valentine to exercise a watching role over tunnel development in 1957, the role of British Railways was ill-defined, but it was assumed that the train ferry shipping service would be diverted to the tunnel, ie the freight wagon service and also the Night Ferry sleeping car train. So far as other through passenger services by the tunnel were concerned, it was expected that there would be interchange between ordinary Southern Region trains and SNCF trains at a new transfer station on the British

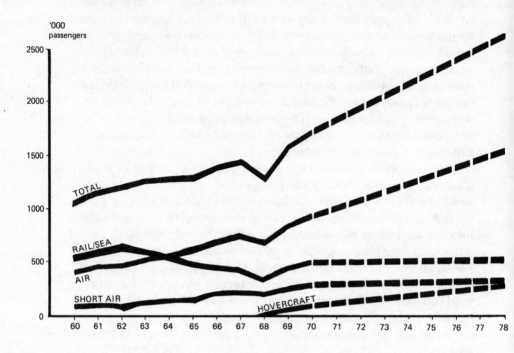

'000
passengers

2500

2000

1500

1000

500

0

TOTAL

RAIL/SEA

AIR

SHORT AIR

HOVERCRAFT

60 61 62 63 64 65 66 67 68 69 70 71 72 73 74 75 76 77 78

Trends in passenger traffic: London-Paris

side. The site for this had been tentatively selected at Westenhanger, six miles west of Folkestone, in the 1960 Study Group report. The Southern Region was not, in 1960, yet electrified between London, Folkestone and Dover, but this was done in 1962, on the third rail 750 volts direct current system. It was contemplated that any through trains would probably change locomotives at the interchange station between BR and the SNCF. There would be no technical problem in having two systems of electrification in the same station; that happens at many frontier stations on the Continent. At that time too, multi-system electric traction units capable of running on ac or dc supplies at different voltages had not yet been developed although they were only just round the corner.

Valentine became chairman of London Transport in 1959, but it may be noted that his interest in the tunnel continued for a long time; he remained a director of the Channel Tunnel Company alongside the BR official representative until 1969, several years after he had retired, with a knighthood, from London Transport. Meanwhile his role in the Study Group had been passed over by Sir Brian to G. W. Quick Smith, who was designated adviser (special projects) and was a member of Robertson's so-called 'general staff'. Quick Smith was a barrister who had had long and distinguished service in the road transport industry, having been secretary of the Road Haulage Association before nationalisation and a senior executive of British Road Services later on. In his Channel Tunnel work he was accompanied by L. B. Marson, designated assistant adviser (how does one give assisting advice?). Marson was a Yorkshireman with lively mind and a great sense of humour, who had come originally from the London & North Eastern Railway into the nationalised British Transport Commission.

Quick Smith and Marson were fortified by the knowledge that Sir Brian Robertson was in favour of the tunnel. Matters changed somewhat in 1962 with the dissolution of the British Transport Commission and its replacement, so far as the railways were concerned, by the British Railways Board, and also by the replacement of Sir Brian Robertson by Dr Richard Beeching in May 1961. Beeching was cooler towards the tunnel, an attitude at chairman level that persisted on BR for a considerable time, covering the periods in office of his successors, Sir Stanley Raymond (in office from 1965 to 1967) and Sir Henry Johnson who followed him. It was not until 1971 when Richard (Lord) Marsh was appointed that British Railways once more had a strongly pro-tunnel chairman, enthusiastic about the prospects of increased international rail business which the project offered.

In the period between the 1964 and 1966 announcements, the civil servants had been concentrating upon the chief outstanding questions covered in AF66; contact with the railways had died away since the days of the Study Group's intensive planning. However, the joint Wilson–Pompidou statement in 1966 accelerated the Ministry of Transport's work. One day in the summer of that year the author's telephone rang and the voice of the BR management development officer said 'Are you willing to take on the Channel Tunnel in addition to your present job?' An answer was required immediately. The reason for such speed was that the chairman (Raymond) had earlier that day been told by the minister, Mrs Castle, that an assistant secretary (Channel Tunnel) had been appointed in the ministry and the name of his opposite number in the British Railways Board was demanded. The chairman had undertaken to give this information to the minister without delay but on returning to his office found that there was actually no such incumbent!

So for me began seven years' work on the railway aspects of the tunnel, always interesting, sometimes encouraging but too often frustrating. At the outset it was made clear that BR's resources available for the tunnel planning were restricted. In each of the principal departments – operating, commercial and engineering – one officer was nominated to whom questions concerned with the tunnel could be referred: he would *not* be available full-time and in fact the time he could spend on tunnel matters would depend entirely on the attitude of his departmental chief. The author's role was meant to be a channel of communication, and co-ordinator of departmental views – a very unsatisfactory state of affairs.

The SNCF position was very different. Tunnel work was directly supervised and stimulated by the deputy director general, M. Roger Hutter; heads of departments participated directly in the meetings which he chaired.

Contacts with the Ministry of Transport soon became necessary and frequent. John Barber, the assistant secretary in charge of the administrative and international planning of the work – able and energetic – was soon joined by an opposite number on the technical side, a retired Royal Engineers officer, Brigadier John Constant. A minor problem arose when Anglo-French meetings required minutes in French; the rank of 'brigadier' in France is that of an a non-commissioned officer, and Constant was anxious that no misunderstanding should arise. It was eventually arranged to translate brigadier as *général de brigade*.

Barber had, very wisely, managed to establish a good personal rapport with his opposite number in the French civil service, Roger

Macé; this meant that Anglo-French arguments, which (inevitably) arose quite often, were conducted without rancour.

The first issue that the author encountered was that of interchange between BR and SNCF trains on the English side of the tunnel. It appeared largely to nullify the attraction of the tunnel – changing trains has always been a commercial handicap, and the extra journey time must reduce the tunnel's attraction still further. In this, fortunately, the BR headquarters operating and commercial officers concurred. But how was through running to be achieved? The alternatives were either enlarging the Southern's loading gauge to UIC standards ('Berne gauge' – roundly 1ft higher and wider than BR, but see the detail allowing for overthrow, and gauging problems in Appendix B) or persuading the Continental railways, not just the SNCF, to accept rolling stock built to the 'train ferry' gauge – ie British overall dimensions but couplings, brakes and buffing gear to UIC standards.

The first option presented formidable problems. There are only two main (or boat train) routes between London and the tunnel portal. The No 1 route, as the Southern Region terms it, includes three very long tunnels – Penge (2,200yd), Polhill (2,611yd) and Sevenoaks (2,453yd). The cost of enlarging all three to Berne gauge would be prohibitive and the prolonged interruption to traffic on the main line would be quite unacceptable.

The No 2 route avoids Penge Tunnel by using the Catford Loop between Brixton and Shortlands, and the Polhill and Sevenoaks tunnels by following the old London, Chatham & Dover secondary route to Ashford, via Otford and Mainstone East. It is however over four miles longer, and slower than the No 1 route because of junctions and curves. However, a gauging exercise on it was carried out by the Southern Region – not in detail, but on a sampling basis – which suggested that obtaining Berne gauge clearance would involve an outlay of £35–£40 millions, for no perceptible advantage to BR. In fact, provision of locomotives and rolling stock by the French would involve payments which would be deducted from the BR proportion of through fares via the tunnel.

The alternative, of the tunnel through trains being built to 'train ferry' gauge, was strongly objected to by the French, though the Night Ferry had operated with through carriages to Paris for many years and the total fleet of ferry wagons in international ownership was numbered in thousands. The French objections were (a) the smaller dimensions reduced passenger accommodation and constituted a commercial disadvantage (b) the need to maintain non-standard spare vehicles at a wide range of centres, both for occasional strengthening and to

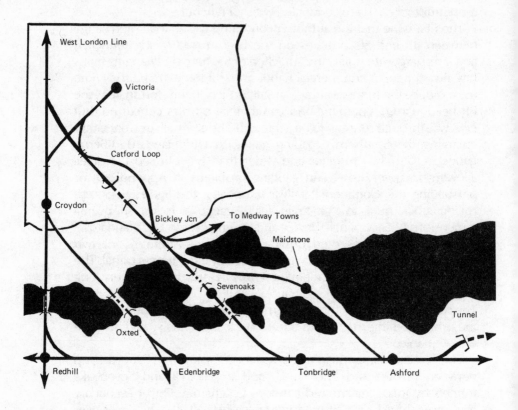

Existing SR routes, Channel Tunnel to London

replace defective units, was unacceptable to the other Continental administrations. The BR representatives could not feel that these were very weighty objections; their main purpose seemed to be the increase of pressure upon BR to accept Berne gauge stock on the Southern Region.

One irrefutable argument was put to the French, that it would be desirable for some tunnel trains, probably of sleeping cars, to originate and terminate elsewhere than in London – probably in Glasgow, Liverpool, Manchester and Birmingham. Such trains in any case would have to be of 'train ferry' gauge, which the French could not deny. For a long time there was something of a deadlock on this question.

A minor Anglo-French difference arose over the question of train lavatories. The British Ministry of Transport's Railway Inspectorate, having consulted the public health experts, insisted that if conventional rolling stock were used it must have the lavatories locked during the passage of the tunnel in order to prevent the discharge of sewage on the track, which the British considered a health hazard. The French disagreed; they pointed out that no such steps were considered necessary in the long Alpine tunnels. This question also was left over, since it seemed likely that by the time the tunnel opened there would be a widespread adoption of retention lavatory tanks of the aircraft type.

The meetings with the SNCF were held alternatively in London and Paris. Sir Stanley Raymond once complained that his officers were bouncing to and from Paris like yo-yos – an exaggeration, but it was certainly the case that there was a lot to discuss and that views often differed widely. The level of fares and freight rates, relations with the operating authority, the probability of on-train customs and immigration formalities, the continuance or otherwise of duty-free sales, the application of different signalling practices and different rule books for staff, were some examples of items that appeared on the agenda papers, quite apart from the crucial question, deferred by mutual consent until a final tunnel decision was reached by the governments, of the type of rolling stock.

The most important immediate issue to be settled by the British planners was the location of the ferry terminal. Westenhanger, provisionally selected at the outset, was replaced by two alternatives – Sellindge, a site alongside the railway and actually lying between the village of that name and Stanford; and Cheriton, north of the suburb of Folkestone bearing that name. Cheriton was adjacent to the planned end of the M20 motorway and much closer to the tunnel portal in Castle Hill below Folkestone.

A substantial public relations exercise was mounted by the ministry to allay fears about any environmental damage from the construction of the terminal. In a consultative document, on which views were invited, the alternative sites were presented as 'packages', each having points in favour and some points of difficulty. The Cheriton package emerged as the most satisfactory from the operational point of view, and aesthetically it involved the least intrusion into the landscape of Kent, being situated on flat ground in the outskirts of Folkestone. The hamlets of Peene, Frogholt and Newington would be affected but othewise little damage would be created. Public opposition to Sellindge was greater than in the case of Cheriton (the latter was in fact favoured by the planners) and it was consequently with satisfaction that the ministry was able to announce the definite choice of Cheriton.

A great deal of head-scratching went on also over the location of the railway station. With Cheriton selected as the ferry terminal site, the tunnel railway would diverge from the Southern line at about milepost 67, east of Saltwood Tunnel, and run straight through the ferry terminal towards Castle Hill. Consideration was given to containing the railway passenger station within the perimeter of the ferry terminal but this idea was quickly found impracticable and a separate passenger station site had to be sought.

It was in fact discussed whether a rail station near the tunnel was needed at all. Would it not be the case that the tunnel trains would make straight for London? If there were a station, would not the customs and immigration authorities insist upon treating it as a frontier point and trying to maintain their traditional insistence that all passengers must file past a static control point? The delays if passagers had to detrain, pass customs and immigration and re-enter the train, would make a nonsense of the tunnel service. But if trains made no stop before London, surely customs and immigration officers could either travel in the train (the best solution) or else carry out their checks at the London terminal precisely as had been done in the case of the Night Ferry for many years.

Finally, it was concluded that, although most trains would not stop at the terminal, it would be impossible not to cater for some amount of local traffic in Kent, either travelling by road to or from the tunnel station or changing there into regular Southern Region trains. Some tunnel trains, though not the majority, would stop in the station for this purpose. So several sites were looked at, and Saltwood, just east of the tunnel of that name, was considered the most suitable, as interchange with the regular Southern Region services would be possible at the junction between tunnel and SR lines.

The main tunnel railway lines would then run in cutting and cut-and-cover tunnel under the central area of the ferry terminal; there would be no stopping there.

A freight terminal was also needed, for some re-marshalling of wagons from the Continent and some transfers to road, although again it was hoped and expected that much trainload traffic would pass to inland clearance depots (ICD). However, customs were expected to insist upon their right to inspect customs seals on vehicles bound for ICDs, close to the point of entry into the United Kingdom, and the freight terminal would have to produce facilities for this. There were two likely sites for a freight yard, one at Sevington immediately east of Ashford, and another at Stanford, almost opposite the Westenhanger racecourse. These options were kept open since action on these and similar issues could not go very far until the governments committed themselves firmly and finally on the vital question of who should build the tunnel, under what conditions, when work should start and when it should be planned to finish.

9 Forward, Slow March

The record of this chapter is depressing since it seems to indicate the unfitness of governments to oversee a great public work. The delays and indecision on the part of both ministers and civil servants were inimical to effective implementation of a project already shown to be technically sound and economically viable. Years were wasted and, naturally, every year the estimated cost increased. Inasmuch as this was due to inflation, it would be matched by increased receipts; but had the tunnel been built between, say 1964 and 1970, the first cost, and hence the interest charges, would have been much lower.

Following the Wilson/Pompidou announcement of 8 July 1966, and the more detailed statements by Mrs Barbara Castle and M. Pisani on 28 October 1966, the way might have seemed clear for rapid progress. That would have been over-optimistic. The governments decided not to accept the scheme put forward by the Channel Tunnel Study Group as it stood, but to invite competitive tenders from 'all interested parties'. The French concept of an exclusive concession, which had taken root on their side of the Channel as long ago as 1875, was not acceptable.

Tenders were eventually received from three groups, in which of course the Channel Tunnel Study Group figured with high hopes of success. Another group was a consortium of banking concerns with particular interests in the eurodollar market – S. G. Warburg & Co, Banque de Paris et des Pays-Bas, and White Weld & Co. The third entrant was a group led by Hill Samuel in London, with the Banque Nationale de Paris and no less than eight other British, French, Italian and American banks.

The initial financing plans were submitted by 5 April 1967, whereupon the groups were invited to put in fuller plans, before July 1967. This was done; but before any decisions was reached a hiccup occurred (this time, and unusually) on the French side. The political upheavals, and the student riots in 1968, made government decision-making difficult for some months; the administrative machine suffered a form of paralysis. This however was not the only reason why the announcement of the group selected to finance and plan the construction of the tunnel, expected in 1968, was delayed. Delay was also due to the fact that examination of the three tenders by ministers and civil servants had not led to a preference for any one proposal – in fact, each tender included some feature or features considered unacceptable. So new guidelines were worked out by the governments jointly and given to the three groups.

Meanwhile the British Government confirmed its commitment to the tunnel, subject to satisfactory completion of a full economic appraisal over the next two years; as an earnest of this intention, the 1968 Transport Act included a provision for setting up a Channel Tunnel Planning Council which would constitute in 'shadow' form the British half of a future Anglo-French Channel Tunnel Operating Authority. It also provided powers for the Minister of Transport to acquire by agreement any land expected to be required for the purposes of the future tunnel terminal. A retired diplomat, Sir Eugene Melville, became chairman of the Planning Council, and some properties in the hamlets expected to be affected by the construction of the Cheriton terminal were purchased, the occupants being given short-term tenancies until the sites should be needed.

By October 1968 Richard Marsh had succeeded Mrs Castle as Minister of Transport and on 23 October he was able to announce that he and M. Chamant, the French Minister of Transport, had completed their consideration of the proposals. On 11 November he told the House of Commons that the choice would be made early in 1969. It was not to be. What happened was that the three competitors were asked to combine and submit a consolidated proposal which would follow guidelines given by the governments. That was not effected until there had been another general election and a change of British government in 1970. Yet another Minister of Transport picked up the reins, this time a Conservative, John Peyton. On 15 July 1970 he told the House of Commons that a new Channel Tunnel Group had been formed, comprising some (by no means all) of the former participants. On the British side, a new Channel Tunnel Company would construct the British half of the project. The old Channel Tunnel Company, dating from 1887, would be renamed Channel Tunnel Investments Ltd and become a founder shareholder in the new Company. On the French side, the construction company was also to be a new creation, the Société Française du Tunnel sous la Manche.

The financing participants on each side were:

British Channel Tunnel Co Ltd. – Founder Shareholders

	% holding
Channel Tunnel Investments Ltd	25.00
The Rio Tinto-Zinc Corporation Ltd	20.00
Morgan Grenfell & Co Ltd	10.50
Robert Fleming & Co Ltd	10.50
Hill Samuel & Co Ltd	10.50
Kleinwort Benson Ltd	10.50
S. G. Warburg & Co Ltd	5.50
British Railways Board	4.74
Morgan Stanley & Co Inc	0.92
The First Boston Corporation	0.92
White Weld & Co Ltd	0.92
	———
	100.00

**Société Française du Tunnel sous la Manche
– Founder Shareholders**

	% holding
Banque Louis Dreyfus	13
Banque Nationale de Paris	8
Banque de Paris et des Pays-Bas	8
Banque de l'Union Européene	8
Compagnie Financière de Suez	13
Compagnie du Nord	13
Credit Commercial de France	8
Credit Lyonnais	8
Société Générale	8
Société Nationale des Chemins de Fer Français	13
	———
	100

There are several points of interest in the composition of the new group. First, of course, is the fading out of the historic Channel Tunnel Company from an active role – and with it, Leo d'Erlanger, who retained the now shadowy chairmanship only until 1974. He died in 1978. Then, the continued participation by the Suez company is notable. But the American interest had dwindled considerably since the days when Technical Studies Inc had had a quarter share in the Study Group. Another feature was the French Railways' 13 per cent participation in the French company, compared with the much smaller 4.74 per cent holding of British Railways in the British company.

But perhaps the most significant change was the appearance of Rio Tinto-Zinc Ltd as a major shareholder on the British side, since RT-Z was also the designated project manager for the British element in the construction work. RT-Z had wide experience in planning and constructing large-scale projects, not merely in mining, its original field, but, for instance, hydro-electric schemes, aluminium smelting and so on. It was obviously well qualified for the task of project management but made it clear that it would participate only as partner and co-investor; it was not interested in consultancy on a fee basis. It formed a new subsidiary company, RT-Z Development Enterprises Ltd for the Channel Tunnel work, controlled by Alistair (later Sir Alistair) Frame, the parent company's chief engineer, who was soon to join the RT-Z main board and eventually to become its chairman. RT-Z's whole-hearted commitment to the tunnel was demonstrated by its assignment to the project of this fast-rising star in its managerial firmament.

On the French side, a more normal consultancy arrangement prevailed with a new group, Situmer, formed out of two firms, Sogelerg and SETEC on the technical side, with SETEC-Economie retained for the economic studies.

RT-ZDE soon arranged to have expert advice on several aspects of tunnel design and construction – Mott, Hay & Anderson, associated with Sir William Halcrow & partners, were the ultimate experts in the field of tunnelling. For the design of the ferry terminal area, obviously a sensitive aspect of environmental grounds, Building Design Partnership was retained; and on economic matters, Coopers & Lybrand were the leading consultants selected.

In March 1971 the governments, acting through the two Ministers of Transport (on the British side John Peyton, on the French side, M. Chamant) announced approval in principle of the consolidated group's proposal. This was to be followed by 'Preliminary Studies' (yet again!); these were carried out energetically by RT-ZDE and Situmer jointly and appeared in a substantial report dated April 1972. This convincing and detailed document updated the previous studies, established a 'reference design' and estimated the cost and time-scale of the project. It put the total cost at £365 millions (which included £33 millions for contingencies). Elaborate forecasting of traffic levels was included and, assuming the tunnel to open in 1980 and to have a 50-year life, the economic benefits were estimated to rise from £45 millions in 1980 to £72.2 millions in 1990 and £101.8 millions in 2000.

The report was sufficiently convincing for the governments to sign, on 20 October 1972, parallel agreements with the companies, collectively known as Agreement No 1, providing for further work to be carried out up to July 1973 at a cost of about £5 millions.

With such formidable forces deployed on the financial, technical and economic aspects it might have seemed that the project had acquired so much momentum as to be unstoppable. That was not to be the case, for several reasons.

Inherently, the organisation was cumbersome, since on each side there were concerned:

> Governments
> Channel Tunnel Companies
> Project Managers
> Railways.

A minimum of eight partners, four from each side of the Channel, was thus involved in most of the important decisions. Arguments were bound to arise. At times the British would close ranks against the French, or vice versa; at other times, British and French railways, for instance, would present a common front against British and French project managers.

Nevertheless, progress was substantial from 1971 onwards; RT-ZDE imparted a needed momentum into the work. Their insistence upon close cost control was also a healthy discipline, as was their keen awareness of the needs of the timetable. It was no doubt natural that they should refuse to take anything done before they arrived as settled, and to re-examine and if necessary modify existing plans in ways that sometimes seemed unnecessary to those who had long been involved with the project. The chief technical and operating officers of the railways were inclined to have, to put it mildly, reservations about having to justify their decisions on purely railway matters to project managers who were not professional railwaymen, and there were occasional moments of testiness.

In general, however, RT-ZDE's re-appraisal of the scheme confirmed the soundness of the earlier planning by the Study Group, though various minor modifications were introduced with beneficial results, including a slight variation in the tunnel's alignment and a revision of the ventilation proposals. The project managers soon became impressed with two economic factors – the potential of the through road lorry traffic which had been expanding fast since 1960 via the roll-on, roll-off ferry ships; and the need to stimulate the through rail traffic. In the latter objective they were at one with the railwaymen who had long disliked the idea of interchange between trains near the English tunnel portal. The way that progress was made here turned out to be unexpected, coming from the French side.

10 The New Railway Project

During a critical period, Channel Tunnel rail planning, like other British Railways activities, was suffering from the effects of high-level pre-occupation with organisational questions – whether or not there should be a chief executive for the railways, and what his powers should be in relation to a board composed mainly of full-time members, all wanting an executive role. Much depended on the outlook of individual chairmen, whether coming from outside the industry like Beeching and Marsh, or from within like Raymond and Johnson.

Dr Beeching had distributed the executive functions among the full-time board members, which was fine except when they were at loggerheads with each other which was sometimes the case! Raymond and Johnson, temperamentally inclined to keep a strong guiding hand on the reins themselves, had delegated in different degrees and at different times to vice-chairmen and chief executives. One of the latter had to be relieved of his duties and another damaged his health through overwork. Johnson was a strong delegator, but only to people from his own stable, who knew his mind and could anticipate his policy line. His attitude towards the tunnel had been sceptical if tolerant; he once told the author 'I don't mind spending a little money and time on the tunnel so long as that does not interfere with the things we really want to do'. This attitude filtered down and a low priority for the tunnel work permeated the major departments, except from time to time when departmental heads like Gerard Fiennes could briefly display some personal interest.

The struggle to obtain adequate resources for Channel Tunnel planning was eased somewhat when the author was permitted to recruit a small team representative of the main railway departments. Some chief departmental officers were inclined to object to this at the outset; but they were mollified by a full explanation to the effect that 'line and staff' organisation principles would be followed. The team would work under the author's chairmanship, identifying problems and alternative strategies that were worth consideration; but each member would consult the departmental chief (or a nominated deputy) to get approval of all major proposals before these were integrated into the planning.

The former assistant civil engineer of the Eastern Region, J. B. Manson, joined the team with a general oversight over the technical aspects of the planning; his expertise was a valuable asset. And a change in atmosphere, if not reflected directly in work patterns, appeared with the arrival in 1971 of a new chairman (Sir) Richard Marsh. His quick mind seized on the tunnel as an opportunity for BR, but he had not quite appreciated the degree of scepticism that had been inherited from previous chairmen and the need for BR to show the same high-level commitment to the tunnel as the SNCF on its side. The board member now involved was David McKenna, a brilliant and charming personality and strong supporter of the tunnel; but he had many other pre-occupations and responsibilities, including chairmanship of the shipping and international business. The big battalions of the main headquarters departments, with recent shifting patterns of organisation and varying relationships with the British Railways Board, had tended to go their own way, as did the Southern Region.

The Southern Region maintained a fairly consistent attitude to the tunnel: it was, that the management had its hands full with running the existing train service and coping with day-to-day problems. Management talent at Waterloo was not to be diverted into planning that might – and in the Region's view probably would not – come to fruition. No pathways could be found for Channel Tunnel trains over and above those now provided for boat trains. There was no spare capacity at the London termini, and plans for reconstructing Victoria did not take account of possible tunnel trains.

McKenna, as a former chairman and general manager at Waterloo, was bound to sympathise with his successors in their difficulties; he knew from personal experience just how many problems existed on the Region and the strain of coping with them. And not he, but only Marsh, was in a position to give a directive that the Southern must rearrange its priorities, if necessary strengthening its management team, in order to participate fully in the Channel Tunnel planning. That did not happen: Waterloo was allowed to go its own way and the headquarters tunnel planning team had to work in isolation that was far from spendid. Certainly from time to time, and 'under the counter', a few friendly and helpful Southern officers would lend an engineering drawing or give some background information over the telephone; but officially something like an iron curtain remained in position.

A new proposition emerged early in 1970; the long-drawn-out and inconclusive discussions between the British and French railways about the three options for the international train services – 'all change' on the British side of the tunnel; 'Berne gauge' and Continental trains to London; or 'ferry gauge' stock on the Continent – suddenly took an unexpected turn. At one of the regular BR/SNCF meetings in Paris the French revealed that they were designing an entirely new type of train service, then named 'Europolitain', later TGV, or 'trains à grande vitesse'.

The object was to attract long-distance travellers away from air and back to the railway and also to stimulate entirely new business. This desire had already led the European railways within the International Union of Railways (UIC) to develop the network of Trans-Europ Express or TEE trains. These had been the brainchild of Dr den Hollander, the president of the Netherlands Railways, who was confident that the railways could compete for high-class business traffic between important centres with a modern type of service, borrowing several features from airline practice. These included compulsory seat reservation, meals at every seat, first-class only accommodation, and frontier formalities streamlined as far as possible. Diesel multiple-unit train seats, reversing without the time-wasting need to attach and detach locomotives, were the original standard; their high acceleration and hill-climbing capacity meant that they could quickly attain the maximum permitted line speed everywhere, while journey time would be cut by all the frontier formalities being carried out while travelling. Station stops would be kept to the shortest possible duration.

TEE had on the whole been a success story, but it worked within sharp limits imposed by the characteristics of existing railway routes – gradients, curves, junctions, and also the conflict with slower-moving traffics. The original concept had had to be modified in various ways – locomotive haulage was needed in some cases with trains dividing for several destinations, as on the Rheingold Express.

The effectiveness of competition with air was thereby being limited to a disturbing extent. Journey times needed to be cut a great deal more, and that meant higher speeds than were possible with the existing infrastructure.

Building an entirely new railway for very high speed was bound to be extremely costly; but the costs tended to be concentrated in the built-up areas around the terminals and in the earthworks and structures needed to maintain acceptable gradients. The Europolitain concept offered savings in cost through using the existing city terminals and tracks in the built-up areas but diverging to a new high-speed route as soon as open country was reached. Here, although curvature had to be of very large radius to permit very high speeds, use of a high power/weight ratio would enable much steeper gradients to be accepted without significant reduction of speed. This meant that tunnels, viaducts and heavy earthworks would usually not be needed. In fact the acceptable ruling gradient was approximately that adopted on motorways, namely 3.5% or 1 in 28. In some cases it was to be expected that construction of a Europolitain line might be twinned, for much of its length, with a new motorway, with an appreciable saving cost.

Speeds of 300km/h (187mph) were planned, with spectacular reductions in journey time. Only high-speed passenger trains would use the new infrastructure, and there would be few, if any, intermediate stations. The timetable should thus be achieved with almost complete punctuality.

The indications, BR was told, were that the French government would look favourably upon this SNCF initiative. For the first Europolitain route to be built, there were two options. One, linking Paris with Lyon, was called 'Paris-Sud-Est'; the other would link Paris with Brussels and (by means of a triangular connection) with the Channel Tunnel; it was called 'Paris-Nord'.

The decision as to priority would depend upon a final commitment by the governments to the Channel Tunnel. Now, the French said, why cannot the British plan a Europolitain line from the tunnel terminal to the outskirts of London? After all, the distance would only be about 60 miles, or less than one-third of the new construction involved in Paris-Nord. Could the British not contemplate such a modest contribution to a splendid joint project? London–Paris journey times might well be 2¾ hours, and London–Brussels 2½ hours. Competition with the airlines must be highly effective.

The British received this proposition with mixed feelings. First of all, obviously, Continental rolling stock and (probably) SNCF locomotives would be running through to London. National prestige would be affronted, and trade union hostility would be aroused. Then, whereas the Nord Region had ample line capacity from the Gare du Nord in Paris as far as the outskirts of the city, the Southern Region was overtaxed by its suburban traffic in the peak hours, and decanting high-speed trains into the middle of commuter services was, in the Southern management's view, impracticable. So the new infrastructure could not begin and end in the outer suburbs as in Paris but would have to penetrate the much larger built-up area, even as far as the terminus, with serious effects upon capital cost.

On the other hand, the attractions of high-speed London–Paris and London–Brussels services were undeniable. Estimates of the traffic to be gained from the notional journey times were very encouraging. The French indicated that unless and until the British had looked into the possibility of Europolitain on their side it would not be profitable to go on discussing the alternative types of train service.

The British team was in a dilemma. It did not seem right to refuse to examine this challenging proposition; on the other hand, at Board level there was not likely to be any marked enthusiasm, to say the least of it. BR's technical resources had been taxed over the recently completed electrification schemes, the demise of steam traction and the problems involved in running-in too many classes of new diesel locomotives and multiple-units. Civil and signal engineering resources were stretched by permanent way and re-signalling demands, not to mention many structures urgently needing replacement.

It was agreed that a feasibility study of possible Europolitain routes would have to be made, but the BR chief civil engineer indicated that he had no resources available for such a task. The only possibility was the use of consultants, and Board authority was obtained to commission such a study from Livesey & Henderson, leading consulting engineers, in June 1970. Their report was received in November of that year. It dealt, as desired, with the possibility of a very high speed railway being constructed between London (Victoria station) or Olympia on the West London line, and the proposed passenger station at Saltwood on the Southern Region, where the tunnel railway diverged from the London–Folkestone main line.

The report considered separately the problems of traversing the built-up area, and the open country. In order to achieve track geometry that would permit the very high speeds postulated, four alternative routes were investigated, leading into the built-up area either at Bickley or South Croydon. All the routes followed the Southern Region alignment from the tunnel zone to west of Ashford (to be bypassed on the south side). Thereafter they diverged as follows:

Route A New alignment via the Maidstone area, crossing the Medway south west of Rochester and continuing alongside the Southern near Bickley, then following the Catford loop.
Route B Diverging from Route A north west of Maidstone, to curve around the North side of Sevenoaks, to reach Bickley.
Route C Leaving the main line at Paddock Wood, and turning north west to aim directly for Bickley, tunnelling under the North Downs.
Route D Following the main line west, past Tonbridge, and then along

the Tonbridge–Redhill alignment as far as Crowhurst, where two alternative routes to South Croydon were plotted, one via Coulsdon and the other following the Oxted line.

The capital costs of the four routes – purely for engineering work, excluding signalling, electric traction equipment, land purchase, or work in the terminals – were, depending on the options:

Route A	£50.4 – £61.2 millions	
Route B	£44.4 – £55.2	,,
Route C	£50.7 – £61.5	,,
Route D	£54.2 – £57.0	,,

The Report was presented to the BR Board who noted it without special interest. The planning team however had to examine its implications very closely.

First of all, it was essential to allow for the fact that these cost estimates, subject to inflation and also, on the consultants' own calculation, to a plus or minus range of accuracy in the order of 20 per cent, were far from comprehensive. Signalling, terminal facilities, electrification and land purchase costs would substantially increase the totals. It was suggested that a 'charm price' of £99 millions should be regarded as the absolute limit: the psychological (even if quite irrational) effect of going into three figures might well cause the project to be ruled out of court straight away.

Secondly, any sufficiently violent environmental objections might well kill any scheme. Routes A B and C went striding across large areas of unspoilt countryside (route A involved a huge viaduct across the Medway valley) and could be expected to arouse fierce opposition. Parliamentary powers would be required and would be bitterly contested.

Only route D appeared to minimise this area of risk, for much of its length constituting in effect only a widening of existing Southern Region lines. New construction over virgin ground could be almost confined to an Ashford by-pass line and a new chord line linking the Tonbridge–Redhill route with the Oxted line.

Planning therefore was concentrated upon a modified version of route D, using the existing Tonbridge–Crowhurst tracks and also those of the Oxted line, with widenings at various places. This seemed to be the most realistic option and it effected some savings over the consultants' original route D which duplicated the Oxted line with new tracks.

There could of course be no commitment at this early stage and the atmosphere only warmed after Marsh became chairman in September 1971. Meanwhile ominous signs were appearing on the cost front. The Southern Region unhesitatingly rejected the consultants' outline plan contained in route D for re-modelling the tracks from South Croydon to Selhurst which would give the Channel Tunnel trains a clear run from the Oxted line to the fast lines between Selhurst and Victoria. They would not consider the tunnel trains using the latter, though in later years the Region has provided a Victoria–Gatwick express service at a 15-minute headway throughout the day and much of the night, with no apparent difficulty since the Victoria resignalling and the track re-alignment north of East Croydon.

In consequence of the Regional attitude it became necessary to envisage a tunnel under the East Croydon complex rejoining not the Brighton main line but the Crystal Palace line at Steatham Hill.

From there an additional pair of tracks was to be provided as far as Clapham Junction whence (again because of the Region's reluctance to contemplate the use of Victoria) the West London line would be followed. This strategy offered some attractions. The former intensive user of the West London by freight trains had greatly diminished and the pre-war passenger services between Willesden Junction and Earl's Court, and Olympia and Edgware Road, as well as that between Olympia and Clapham Junction, had all disappeared, largely owing to bomb damage and changing travel patterns. Olympia was now only a shadow of its pre-war self, when it was then named Kensington (Addison Road); its only regular daily passenger service was one in the morning and evening to and from Clapham Junction mainly for the benefit of workers in the Post Office Savings Bank (not publicly advertised), together with a shuttle service by London Transport to and from Earl's Court during exhibitions at Olympia. Since 1965 Olympia had become a Motorail terminal but this only involved a very few departures at night, and arrivals in the early morning. There was thus ample line capacity for Channel Tunnel trains from Clapham Junction, though the bad condition of the Chelsea rail bridge over the Thames imposed a severe speed restriction which was something of a handicap. But connections with the Western and London Midland Regions were excellent and through trains from the Continent, or interchange at some point on the West London line, could be contemplated.

Walk-outs suggested that a larger and better site than Olympia could be found further north, near White City. There was ample width for a terminal, and connections were good with both the road network

(Westway) and London Transport (Central and Hammersmith & City lines). So planning on the basis of using White City as the Channel Tunnel terminal was intensified.

At the same time, sight was not lost of the need to keep other planning options open. For a time the ministry was sceptical about the justification for the project: the author was in fact rebuked by the Minister of Transport, John Peyton, for publicly advocating a high-speed link as the best means of exploiting the tunnel's competitive power. It was therefore recognised that a fall-back strategy must be retained, and this would almost certainly involve pushing the SNCF, on the one hand, into accepting 'ferry gauge' rolling stock, and the Southern Region on the other into a more flexible position about the use of its tracks by tunnel trains. For instance, if one examined the Southern's working timetable, in addition to the pathways for advertised boat trains to Folkestone and Dover, there were a number of additional pathways ('Q' paths) which could be taken up in case of late running caused by delays at sea or at the ports. Together these pathways constituted the basis for quite a respectable tunnel service. What then was the difficulty?

The Region replied that many pathways were alternatives and not complementary, since platforming at Victoria could not be arranged for trains running in all these timings. This of course could suggest either the use of the West London line or extension at Victoria, which could possibly be managed by a two-level solution, with Channel Tunnel trains using platforms on a raft above the existing tracks. The falling gradient into the station from the Grosvenor Road bridge was a favourable factor in this solution.

An element in the Passenger Department at BR headquarters remained opposed to the White City terminal and strongly preferred Victoria. Eventually a provisional working compromise was reached. Both should be utilised: the White City for some of the daytime trains and of course all the through trains from the provinces, sleeping car trains and (probably) some Motorail services; Victoria for a number of London–Paris and London–Brussels high speed trains.

All such planning however remained merely provisional until the government made up its mind about the international rail services and the investment required by the alternative strategies on offer.

11 Problems of a railway in tunnel

Laying aside the arguments between the British and French railways over the type of the international train services – to be resolved when and if the British Government made up its mind about funding an independent and high-speed rail link from London to the tunnel – work was concentrated for a time upon the interesting problems of railway operation within the tunnel itself. Obviously different considerations would apply to the ferry trains and the through trains – to begin with, the aerodynamic resistance would not be the same owing to the different cross-sections. Different speeds were envisaged for the through passenger trains, the freight trains and the ferry trains.

An early question was whether the ferry trains should be open, as in the case of some car-ferry services through Alpine tunnels, with the passengers remaining in their cars, or enclosed. Both on safety and amenity grounds, the latter solution had advantages. It would permit car passengers to stretch their legs, visit toilets and obtain drinks and snacks from vending machines, possibly also enjoying the facility of duty-free sales, although this would depend on the attitude of the customs authorities. (From the railway point of view the latter point was of minor importance, since the receipts would accrue to the operating authority and not to them.) The covered ferry vehicles would accommodate, as well as cars, caravans, buses, and smaller commercial freight vehicles. However, the large articulated heavy goods vehicles would be carried on open flat wagons, with their drivers remaining in their cabs.

With the double-deck trains having a capacity of 268 cars, a maximum train length of 750 metres, with a loaded weight of 1,200 tonnes, it was calculated that locomotives of 6,000 kilowatts (or roughly 8,000 horsepower) would be needed to provide the required performance.

It was expected that the two railways, with their accumulated expertise, would be the contractors to the authority for the maintenance of the fixed railway equipment in the tunnel. There would be no doubt that, on the British side at any rate, the standards would have to be approved by the Railway Inspectorate in the Department of the Environment's Transport Ministry. It was agreed that the use of ballast should be ruled out and that the track should be laid on a continuous concrete slab. This would reduce maintenance costs in return for a higher initial outlay and assist in keeping a near-perfect alignment and 'top' on the running rails.

Maintenance was planned to be carried out at night, when the tunnel would not be closed but single-line working would be introduced between crossovers, allowing possessions for working teams. A vehicle travelling on rail at 10km/h (6mph) would be the chief means of track and signal inspections, with two-hour possessions, while longer possessions would be needed for remedial work.

The service tunnel would of course be intersected by the rail crossovers, if its alignment was always parallel to that of the running tunnels. In principle, level crossings at such points would be undesirable since the times when the service tunnel would be most intensively used for maintenance would also be the times when the crossovers would be in use for single-line working. Consideration was therefore given to taking the service tunnel over or under the crossover tunnels. Level crossings however remained likely.

The signalling would have to take account of differences in BR and SNCF practice – for instance, the SNCF does not use overlaps though signals are sited 100 metres to the rear of a fouling point such as a converging junction. On the whole, four-aspect colour-light signals could be considered adequate for the three speed bands envisaged for the different classes of train – 160km/h (100mph) for the international express trains, 140km/h (87mph) for the ferry trains and secondary passenger trains, and 120km/h (75mph) for freight trains, which would be largely of Freightliner type. Signal spacing, having regard to the close headways of 2½ minutes envisaged, would be different on the falling and rising grades, in order to keep headways as even as possible. The question of a more sophisticated type of cab signalling being required, as compared with standard British and French practice, was left over for final agreement later. Continuous speed control was certainly desirable in order that the close headways envisaged could be maintained. The great weight of the ferry trains would mean that a signal check at the foot of the long 1 in 100 gradients leading out of the tunnel could reduce the acceleration so much as to impose reaction on following trains to an unacceptable extent.

The design of the ferry terminals would be a key factor in determining the throughput capacity, since the loading and unloading times were critical for any planning of headways and services. The Transport and Road Research Laboratory at Crowthorne in Berkshire

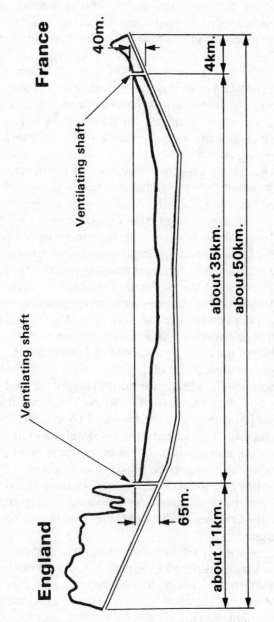

Longitudinal tunnel section

constructed a mock-up ferry train and arranged for a number of private car drivers to pass through a system roughly corresponding to the proposed access platforms for loading and unloading. The results were very satisfactory. A double-deck train could be completely unloaded and re-loaded in a 'follow-my-leader' operation, in around 10 minutes.

Problems affecting both ferry trains and through trains were aerodynamic resistance, ventilation, drainage and temperature. The piston effect of the ferry trains, with their large cross-section, would obviously be much greater than in the case of the through trains, and this would increase the power requirements of the locomotives although assisting ventilation.

The passage of a high-speed train would create cross-winds of considerable velocity, via the adits, both in the service tunnel and in the opposite running tunnel. Accordingly, during maintenance periods when staff are working on the ground speeds might have to be restricted. During single-line working, ferry trains would be limited to 100km/h (60mph) with 70km/h (45mph) over the tunnel cross-overs.

Originally it had been suggested that the piston effect of trains in the tunnel would solve the problem of ventilation. William Low had argued long ago that twin tunnels with cross-passages would, on the basis of mining practice, also achieve this. More modern studies suggested that a degree of artificial ventilation would need to be applied to the central area of the tunnel and this was to be catered for by trunking in the service tunnel.

The character and location of the control centre for the tunnel was a major issue, since three authorities – BR, the SNCF, and the operating authority – would need to be in close and continuous liaison. Most railway international frontier controls are simple, and in land frontier regions bi-lingual staff are usually available in any number required. Whether the control should be situated in England or France was obviously a matter of some delicacy; one solution discussed was for the main centre to be supplemented by a subsidiary centre on the other side of the Channel which could take over the control function in case of emergency.

A continuous speech link between the control and the drivers of all tunnel trains was considered essential. A link is provided on BR at signals from which a telephone to a signalbox is available; and on the London Underground by continuous contact wires on the tunnel wall, to which a portable telephone can be clipped. But full contact with trains in motion would be needed in the Channel Tunnel; the main technical problem was the addressing and recognition of individual

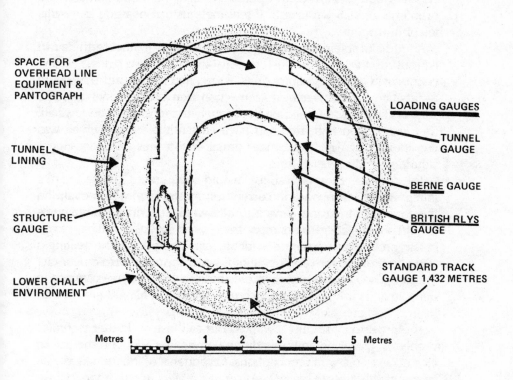

SPACE FOR
OVERHEAD LINE
EQUIPMENT &
PANTOGRAPH

LOADING GAUGES

TUNNEL
GAUGE

TUNNEL
LINING

BERNE GAUGE

STRUCTURE
GAUGE

BRITISH RLYS
GAUGE

LOWER CHALK
ENVIRONMENT

STANDARD TRACK
GAUGE 1.432 METRES

Metres 1 0 1 2 3 4 5 Metres

*Diagram of Tunnel Loading and Structure Gauges (*British Rail*)* trains, which might be overcome by a train describer system. Such a system would also offer the possibility of continuous contact with maintenance staff working in the tunnel, instead of using individual telephone points.

Another problem considered was in relation to the earth return current of the locomotives, due to the high resistance, or low conductivity, of the concrete rings of the tunnel structure. This would necessitate careful electrical connection between the rails and the surrounding Lower Chalk, via the tunnel structure. The Lower Chalk itself was tested and found to have an earth resistivity which was considerably lower than the normal resistivities for the Home Counties.

Telecommunications cables would require separate earths independent of the traction current return earthing, which could be provided in pits in the service tunnel when the drainage channels should provide adequate dampness.

The probable seepage of water was considered as also creating a slight humidity in the running tunnels which would help to counteract the slow rise in temperature to be expected from the passage of the trains, mainly from braking which transforms mechanical energy into heat.

The provisional solutions worked out to these and other technical problems, usually in amicable discussion between BR and the SNCF, cleared away many misconceptions. Opponents of the tunnel, as well as some without firm convictions on the subject, were prone to raise questions that needed patient study before a reassuring answer could be given.

One of the most persistent fears expressed by critics related to safety in the tunnel. This was based upon emotion rather than reason because, if the annual reports of the Railway Inspectorate in the Department of Transport were analysed it is clear that most of the circumstances surrounding or significantly contributing to train accidents would be absent in the tunnel. To list some of the most important factors: the track was to be continuously welded without joints and also laid on a continuous concrete slab, in a virtually stable temperature, so that track distortion would be practically impossible. There would be no route junctions; trains would not be stopping at intermediate stations; loose-coupled four-wheeled wagons would not be allowed; there would be no public highway level crossings.

To experienced railway operators, all the tunnel problems appeared readily capable of solution insofar as their nature was technical or operational. The difficult questions were those of administration, finance, and politics.

12 Strange Bedfellows in Opposition

Opposition to the tunnel was divided between that directed at the tunnel itself and that against the rail link. In the former case, it was predictable that strong resistance would be experienced from vested interests that feared damage. First and foremost came the shipping lines based on Dover and Folkestone, the Dover Harbour Board, and the National Union of Seamen. The shipping lines comprised chiefly those of British Rail (later to be part of the Sealink consortium) and the Townsend Thoresen interest. As the British Railways Board officially supported the tunnel, its shipping side had to tread delicately in opposition. No such difficulty afflicted the Townsend organisation, whose managing director, Keith Wickenden, was energetic in mobilising resources against the tunnel.

The Dover Harbour Board also mounted a strong campaign, alleging damage to employment and business in the town. It went further and sought to discredit the financial estimates of the tunnel promoters by obtaining 'independent' studies of the project's prospects. The term 'independent' is interesting, since the studies were commissioned with the sole object of damaging the outlook for the tunnel; they would not have been used had they been favourable to it. Against that, the studies commissioned by the tunnel promoters were truly 'independent', in the sense that had they been pessimistic the tunnel might not have gone ahead!

Folkestone showed a more mixed attitude. The port would suffer loss of business, but it was owned by British Rail and was not in a position to agitate separately from the town, where the council saw employment flowing from the Cheriton terminal as an offset to the losses in the port. During the construction period moreover Folkestone, and to some extent Dover, would experience an inflow of resources in men, materials and money spent in the towns.

Ashford's reaction was different from that of Dover. The large railway workshops, which specialised in wagon construction and repair, were threatened with closure. The works were well situated for the construction of at least one-half of the ferry vehicles, and for the subsequent maintenance of the whole fleet: no equivalent workshop capacity existed near Calais on the French side. So the tunnel offered the prospect of a continued workload for an establishment which had always been important in the business life of the town.

In France, while Calais was certainly on balance hostile, though the creation of the ferry terminal where would create employment to offset the loss of shipping activity, the Département du Nord was strongly in favour of the tunnel. This region of France was not sharing in the prosperity of the rest of the country, and improved international transport links were seen as a major factor in improving matters. The city fathers of Lille were particularly strong supporters. The SNCF's proposed high-speed rail link with Paris, unlike the much-criticised BR link with London, was strongly welcomed.

The National Union of Seamen was unequivocal in its opposition, naturally enough. But the shipping lines and the union were both caught to some extent by their contention that a fixed link was inferior to a flexible (ie shipping) link. If the latter was truly flexible, resources could be easily transferred to other routes outside the traffic shadow of the tunnel. The tunnel promoters had always accepted that, while only minimal shipping services were likely to continue at Dover and Folkestone, Newhaven–Dieppe would do better while Southampton or Portsmouth to Le Havre, St Malo and Cherbourg would be little affected. Dover–Ostend would continue but with some reduction in frequency, while Harwich–Hook of Holland would scarcely be affected at all. And there would be five or six years at least during which the redeployment of ships and crews could be planned by this 'flexible' industry.

The trade union movement was divided. Against the strong hostility of the seamen's union, the railway unions welcomed the tunnel, apart from some suspicions that the government and the promoters might not ensure that sufficient maintenance and operation work was strictly preserved for British railwaymen. The Transport & General Workers Union building section could see attractive employment prospects from the tunnel construction work. On the other hand the road haulage section tended to be hostile. First came the prospect of a transfer of international freight from road to rail especially if Freightliner type services of containers to a wide range of Continental destinations were established. There was also a more subtle influence. On the roll-on, roll-off ships, long-distance lorry and coach drivers were provided with reserved accommodation and facilities separate from those of ordinary passengers. There was some competition in the standard of such amenities provided between the shipping lines, to induce the lorry drivers and their owners to patronise a particular service. This presumably would no longer be the case in the tunnel, but then the new tunnel transit will be about 30 minutes against a sea crossing by ship of roundly 1½ hours.

Furthermore, a period of a couple of hours or more without driving might rank as a statutory rest period in a driver's schedule; the half-hour transit of the tunnel would not qualify in the same way.

Opposition to the terminal at Cheriton dwindled after its selection was announced. A few residents of Peene, Newington or Frogholt thought that it was worth while continuing to campaign but the steam had gone out of their campaign since a number had sold their properties to the Department of the Environment and remained as tenants subject to notice.

Some naturalists objected to the site of the tunnel portal on the ground that it would disturb the unique habitat of the Spider Orchid. This potential misfortune aroused the anger of some nature conservationists to a degree only possible in Britain. That tiny herb is difficult to distinguish from the Bee Orchid and some floras do not even list it as a separate species. Any colonies of it could however be lifted with due care and replaced nearby in the downland which is its home.

Other opposition to the tunnel was based on alarmist suggestions which were exploited by interested parties. It was suggested that rabid dogs might infiltrate this country from the Continent. The principle that a single point of entry control must be much more effective than those at a range of small Channel ports was ignored. The possibility of fire or sabotage was constantly raised, as though ships were not already vulnerable to a bomb in the boot of a car, or a fire in the engine room. Much patience had to be employed in explaining that, while 100% safety is never attainable on the road, in the air, or on the rail, the precautions designed for the tunnel should make it an exceptionally safe stretch of railway for the reasons explained at the end of the previous chapter.

Lastly, there was the simple xenophobia, the dislike of strengthening links with Europe, that was to distinguish much of the debate over Britain's entry into the Common Market; it spilled over on to the tunnel.

Opposition to the rail link showed many curious features. It divided into hostility to the actual construction, and fears about the effect of high-speed trains. Construction of the rail link was subject to the parliamentary private bill procedure. But in addition to this, the Department of the Evironment asked BR to engage in public consultation of the kind associated with motorway construction, before finally deciding the route of the link. Accordingly in January 1974 BR issued a 'Document for Consultation' showing in twelve fold-in plans the route from White City to Cheriton with alternative alignments in two sections – the by-pass, in the environmentally sensitive area of the North Downs between Woldingham and Edenbridge, avoiding Oxted town and Oxted Tunnel; and the Ashford by-pass to the south of that town.

Local feeling ran high here. But, interestingly, one local residents' association produced an alternative to the route favoured by the Surrey County Council – and this alternative also seemed preferable to BR. All kinds of rumours were started: one was, that BR had relied upon maps 30 years out of date in preparing the drawings for the consultation document. In reply, BR pointed out that up-to-date aerial surveys had been used for the large-scale plans. To show the general alignment of the proposed new tracks on a scale convenient for the consultation document, however, the latest Ordnance Survey maps had to be obtained and used. BR appreciated that all Ordnance Survey maps are to some extent out of date by the time they are published, but (they wrote rather plaintively) 'it was not thought that anyone would imagine that actual scheme planning would be done on the basis of such maps'.

The total number of enquiries received in correspondence and dealt with exceeded 3,000. BR also produced a booklet, of which 10,000 copies were printed, entitled *Your Property and the Rail Link*; copies were given to all enquirers and also sent to any landowner affected. It set out landowners' rights and explained that anyone whose property was required for the rail link would, under the Land Compensation Act, 1973, receive the full market value, disregarding any depreciation caused by the announcement of the rail link – including, in suitable cases, 'home loss' payments to cover the costs arising from disturbance.

Even so, at public meetings hostility was shown which it was hard to defuse even by patient explanations and dispelling of misconceptions. Some MPs, though holding office within a government which had decided that the rail link was to be built, were inclined to attract votes by attacking BR's proposals within their constituency, instead of helping to explain them and allay unnecessary fears.

One cause of alarm arose from a report that high-speed trains in Japan were environmentally damaging on account of excessive noise. It was even stated that those living close to the Shinkansen lines in Japan were abnormally subject to psychiatric disorders. BR had patiently to explain that conditions here would be totally different.

Some of the Shinkansen network had been built on low steel viaducts over which the track was not ballasted, so that train noise was excessive. No such construction was contemplated here. Japanese houses built very close to the tracks were of much more flimsy construction than in Britain – another factor.

The fear of excessive noise was however exploited by tunnel opponents with such energy that BR produced a booklet *Noise and the Channel Tunnel Rail Link* in 1974, explaining that tests of noise levels had been carried out with the electric express 100mph trains on the Euston main line and also on the 90mph Southern Region which showed no cause for concern. The decibel (dbA) level would only rise by about 7½ dbA if speeds of 125mph were reached on the rail link; and the noise – unlike that of motorway – would not be continuous. The booklet explained that 'people living near the Channel Tunnel rail link . . . will hear virtually no railway noise at all for about 55 minutes in any hour, with noise levels slightly higher than the continuous motorway level at the worst point for about 80 seconds in any hour'. In addition, BR promised to erect sound barriers alongside the track at sensitive places, though 36 miles out of the 75 being either in tunnel or cutting would greatly reduce the number of such spots. Out of the total 39 miles to be built on level ground, only six miles would be in built-up areas. Even so, noise continued to be a stick with which to beat both the rail link and the tunnel.

Some curious opposition was experienced from local authorities. Both Kent and Surrey County Councils were willing to accept the rail link as inevitable but were concerned, reasonably enough, to agree the alignment with BR. It was the London boroughs and the Greater London Council that showed the most irrational opposition, though BR's planners could take comfort from the fact that any views expressed by the GLC's officials would be anathema to any borough affected, and vice versa. The GLC excelled itself over the choice of terminal. Objecting to the White City (on no very specific grounds) it asked BR to consider ten other sites. The BR team expressed willingness to co-operate and asked exactly what the GLC had in mind. The GLC eventually supplied a map of South London with a collection of dots indicating suggested terminal sites. Several of these seemed to have no rationale whatever; they were not even sited on or near a railway line and had no Underground links. Formidable destruction of house property would be required to reach most of them. The only serious proposal was the Surrey Docks area, which from a transport angle had many disadvantages. Eventually the GLC dropped out of the debate.

But the two very different boroughs of Hammersmith and Kensington also became hostile. Kensington objected because the West London line, which before the war had carried a heavy traffic but recently had been little-used, was to regain some of its former importance. The peace of the Royal Borough must be preserved, it seemed, at all costs. Yet BR pointed out that Kensington had not previously objected to the puffing and clanking of frequent freight trains in steam days; the future swift passage of quiet electric expresses would be far less intrusive.

Hammersmith, privately expressing derision of the GLC's views, was opposed to BR's intended acquisition of two acres of land representing only about 10 per cent of an area which it had earmarked for later housing development. BR offered to release to Hammersmith other land for housing which would more than compensate for this. BR was criticised for retaining a site at Barlby Road in North Kensington where the authority had had its eye upon some BR land it would like to acquire for housing; it was needed for the terminal but, as BR had never declared the site available it was not 'lost', as the critics claimed.

Finally, the Royal Institute of British Architects declared its opposition to the rail link on general grounds, including the absurd charge of planning on the basis of out-of-date maps. BR offered to explain its approach and in return the RIBA invited the BR planners to a working dinner in the Institute's penthouse suite at the top of its Portland Place building. After the architects had voiced their opinions and BR's explanations had been given, the author could not refrain from inviting the president to come to the window and observe the panorama of London; it was gently suggested that architects who had defaced that former splendid skyline with that dreadful series of slabs and up-ended matchboxes were hardly the best people to criticise a transport project that would be useful and relatively inconspicious.

On the whole, the tunnel enjoyed (if that be the word) a bad press. This was partly a reflection of the complex relationships which government policies had made inevitable, and which were difficult to explain shortly to newspaper readers. Over-simplification and mis-statements abounded in press reports. Comment tended to be hostile, on the journalistic principle that attack makes better copy than defence. Even the stately *Times* and the wiseacre *Economist* lost few opportunities of denigrating the project, partly through the prominence given in news items to actions and speeches by tunnel opponents, partly by editorial comment.

The Times had at any rate the virtue of a consistent editorial policy, hostile to the tunnel. *The Economist* however showed some strange changes of attitude. On 27 May 1972 it gave the tunnel cautious support, writing that the RT–ZDE/SITUMER studies had 'shown the tunnel project may be much broader than initially expected, but that the extra revenue could justify the cost'. A year later a silly piece entitled 'Blancmange in La Manche' described the scheme as 'sadly close to fantasy'; it repeated almost every claim, however wild, advanced by the opposition. On 13 April 1974 it referred to 'BR's muddle about the rail link to the terminal', totally ignoring the fact that BR was working within strict remits given by the government. On 30 November 1974 it argued that the government ought to scrap the rail link and go for either a road bridge or a combined bridge-and-tunnel scheme – sublimely ignoring the cost estimates. Yet, astonishingly, on 25 January 1975, under the headline 'Foiled Again, Boney!' it regretted the cancellation: 'whatever the economic argument for abandoning the Channel Tunnel, the decision was taken in the wrong way, and was pretty certainly wrong in itself'. Rarely indeed did *The Economist* approve of anything or anybody!

The bad press may have been partly due to the fact that neither the great city institutions that backed the Channel Tunnel Company nor Rio Tinto-Zinc normally needed to engage in much publicity – in fact, they preferred to avoid it. RT-Z's work was not usually in direct contact with the public: any publicity it required was often of a defensive character rather than a hard sell. The brochures and hand-outs of the tunnel promoters were factual and useful, but not suitable for popular consumption. Television was not employed to help explain graphically and convincingly to the viewers that the tunnel would be a great national asset and that the fears or the agitation of opponents were mostly groundless. So, for the most part, the great British public retained its original prejudices and these were reflected in the often irrational attitudes of members of parliament.

13 'The Slippery Slope'

British Rail was obliged to work very fast in order to meet the sudden requirement to turn a feasibility study, with merely consultants' notional cost estimates, into a firm proposal with detailed plans and costings. It was only in September 1973 that the government, abandoning its previous objection to the proposal, announced that 'a high quality rail link between the tunnel and London, with provision for through services to provincial centres, is essential to the success of the project'. A BR private bill would have to be deposited in November 1974, to receive the royal assent in July 1975, before the final agreement on the tunnel, providing for completion of the project, could be signed in accordance with the contractual arrangements between the governments and the companies.

This imposed a very tight time-scale. Petitions against the BR bill could be lodged by objectors, and these would be heard before a parliamentary select committee under a semi-judicial procedure; witnesses could be called and petitioners could be represented by counsel if they wished. Even before the bill could be deposited, parliamentary plans covering every detail of the route and the works proposed would have to be prepared, together with 'books of reference' showing the ownership of every piece of land, however small, which might have to be acquired.

It was three-quarters of a century since a stretch of new main line railway as long as this had had to be authorised through the private bill procedure. The task would strain the resources of the engineering, estate and legal departments. There was a further complication, an organisational one. The BR Board had accepted the recommendation of management consultants that there should be a clear distinction between the Board's 'corporate' activities, and the various 'businesses' – railways, shipping, hotels and property – which it controlled.

This concept, drawn from the practice of the commercial and industrial world, was unsuited to the railway industry, with its complex inter-dependence and inevitable cross-subsidisation of activities. An obvious fallacy was considering BR as a conglomerate. A true conglomerate is free to expand or contract its subsidiary activities and is not dominated by the results of any one of them. But BR has never been a true conglomerate (since the privatisation policy of the Conservative Government it is in no sense at all a conglomerate). Even in 1972 it was utterly dominated by the railway business which, including its ancillary workshops, accounted for 90 per cent of its gross receipts. The whole undertaking stood or fell by the results of the railway.

So the corporate and business distinction was essentially unrealistic. But this artificial concept had led to the Channel Tunnel planning being treated as a corporate activity, outside the railway business, until suddenly it was realised that the total railway resources must be involved, and very quickly, if the deadline for the new rail link was to be met.

Normally, the substantial drawing office of a Region would have been employed. However, the Southern Region was passing through a very difficult period on both the engineering and the operating sides, and staff were heavily taxed by its day-to-day problems. There was probably some relief among the Waterloo management that an independent fixed link had been approved, which would keep Channel Tunnel trains clear of Southern Region metals. In consequence, it was decided that the engineering design should be carried out by a newly recruited team at BR headquarters responsible to the chief civil engineer and (ultimately) to the chief executive (railways). Unfortunately the corporate and business distinction was perpetuated, in that no formal liaison was provided between the chief civil engineer's Channel Tunnel drawing office and the Channel Tunnel department.

Both the engineering draughtsmen and the estate surveyors (as well as the lawyers) were desperately hard pressed by the demands of the timetable. Matters were not eased by the pressure from the Department of the Environment to conciliate public opinion in ways that were often bound to involve extra construction cost and delay – for instance, sinking the line in cutting to minimise noise and appease the scaremongers. Alternative alignments suggested by local authorities, pressure groups and even individuals had to be given serious and time-consuming consideration.

The length of tunnel proposed grew steadily, partly from amenity reasons, in built-up areas; the condition of the Chelsea railway bridge over the Thames gave rise to a proposal to tunnel under the river to reach the West London line. It was even suggested that the length of tunnel on the rail link would soon be greater than that of the Channel Tunnel itself. One BR man cried in despair: 'Why don't they just join the London Underground with the Paris Metro?'.

The Department of the Environment's concern about amenities was nevertheless understandable. There could be a domino effect if opposition to the BR bill in parliament proved effective, as it could react on the main legislative timetable and the signing of Agreement No 3. So BR was urged to conciliate objectors as far as possible by following in effect motorway construction procedures (apart from a full public enquiry), namely by offering a choice of routes and, so far as practicable, giving weight to public preferences in making the final selection.

This greatly increased the workload on the design team. Alternative plans had to be prepared at a number of sensitive places. In addition, BR departmental officers wanted to consider the effects of alternative lay-outs from the angle of operating efficiency. A diagram of route options, produced by the chief civil engineer's drawing office, resembled a huge jigsaw puzzle.

So BR had the worst of both worlds – the stringent procedural requirements of private bill legislation, and the public consultation methods employed for motorways.

Internally, and understandably, the railway technical departments resolved to implement to the fullest extent the government's decision that the rail link must be 'high quality'. For instance, concrete slab track in place of ballasted track, identical with that to be used in the tunnel, was favoured. A continuous roadway along the open-air sections, to provide easy access for motor vehicles involved in maintenance of track, signalling or overhead line equipment, was allowed for.

Where alternative routes had to be considered, the BR designers had much less freedom than their motorway counterparts, who could accept gradients as steep as 1 in 28. These, incidentally, could be used by the SNCF planners working on Paris–Nord, since freight trains would be excluded from the high-speed line and use the normal route. In England the restricted loading gauge would exclude Continental freight wagons from normal routes but it had been conceded at a meeting with the SNCF that through freight from the Continent could use the high-speed link, thus imposing limitations on the severity of gradients that could be accepted which in turn meant increasing the length of line in tunnel.

As the work progressed, it became obvious that a dangerous situation was developing. The author wrote a paper entitled *The Slippery Slope*, pointing out the possible cumulative effects of the pressure on the Channel Tunnel department and the chief civil engineer's drawing office. It was agreed that an up-to-date re-costing of the whole project was needed, before the design could be completely finalised.

Re-costing was undertaken with, it must be said, a strong inclination towards caution, and seeking protection from any subsequent criticism of under-estimation. The result was alarming – £373 millions compared with the original feasibility study price of £123 million. An immediate inquiry was set up, on the instructions of BR's chairman. But the new figure had to be accepted if full allowance was made for: traction and rolling stock investment, amenity protection, inflation, higher technical standards, extra work, especially tunnelling, and terminal buildings above platform level (previously excluded).

The increase in cost had to be reported to the Department of the Environment, where it caused consternation. It was of course used by the press as a stick with which to beat both the Channel Tunnel and BR, often unfairly. The minister soon made up his mind that the cost of the link was unacceptable – though the French saw no difficulty in providing the SNCF with funds to build Paris–Nord. He announced this in the House of Commons on 20 January 1975.

Not enough notice was taken of the fact that other planning options had always existed and that they could rapidly be worked up into lower-cost solutions. Press reports implied that the government had directed BR to find a cheaper solution: in fact, these solutions had always been there and only the government's earlier decision had led to their being laid aside.

But other and even more powerful forces were at work to produce a crisis in the tunnel planning. Their cumulative effect was to be disastrous.

14 Abandonment!

Although in the press the escalation in cost of constructing the high-speed, high-capacity rail link from London to the tunnel was generally regarded as the chief reason for the Labour Government's sudden abandonment of the project in January 1975, the real reasons lay in the British political system. There was certainly a groundswell of opposition, not entirely confined to back-bench Labour MPs, which led to continued sniping. But the most important single factor was the disruption to the progress of the Channel Tunnel bill through parliament caused by two general elections in quick succession, coming at a critical time. On 28 February 1974 Labour had been returned, but with a majority of only six seats over the Conservatives, and even polling fewer votes, 37.2 per cent of those cast against the Conservatives' 38.1 per cent. Since the Liberals and minor parties had 37 seats it was in effect a hung parliament, with Labour having to purchase Liberal support to carry any measure. On 10 October 1974 a second general election took place which gave Labour 319 seats against the Conservatives' 276, and a tiny overall majority over all parties, so that (with much activity on the part of the government whips) parliamentary business could proceed more or less normally. After the smooth passage of the preliminary Channel Tunnel (Initial Finance) Bill on 13 November 1973, the main Channel Tunnel bill had been introduced on 20 November 1973, and the second reading had been approved by 203 votes to 185. The committee stage was well advanced at the dissolution of parliament in February 1974 when the bill automatically lapsed.

The bill was reintroduced on 10 April 1974 and was given a second reading on 30 April by a comfortable majority of 287 votes to 63. When parliament was dissolved in September the bill had completed its select committee and standing committee stages.

After the general election on 10 October 1974 work on the bill resumed (by means of a procedural motion, approved by 168 votes to 115) at the stage it had already reached, namely that of report. Meanwhile, although the parliamentary machinery seemed to be grinding on despite these interruptions, there were already some rather ominous hints of back-tracking. During the second reading debate on 30 April the Minister for Transport, Fred Mulley, had announced that it was intended to appoint a small high-powered group of independent advisers to review the economic and financial aspects of the project (yet again!) before the decision to go to phase III (ie the final agreement with the companies enabling construction contracts to be placed) would be put to parliament. Sir Alec Cairncross, a former head of the government's economic services, was appointed chairman of the advisory group. Then, on 26 November and while the Cairncross group was working but had not yet reported the Secretary of State, Anthony Crosland, announced in the House of Commons that the timetable for final decisions on the project could not be adhered to because of the need for British Rail 'to examine a less expensive rail link between Cheriton and London'. He said it was 'out of the question' that the Government should approve or finance an investment of £373 million pounds on the rail link.

Anti-tunnellers in the Labour Party were reinforced. In *The Castle Diaries,* Mrs Barbara Castle, who as Minister of Transport in 1966 had announced with apparent pleasure an Anglo-French agreement to go ahead with the tunnel, noted on 14 November 1974 that there were 'mutterings about the Channel Tunnel bill which clearly no one but Tony Crosland wants'. A week later, she wrote 'In my absence Cabinet agreed to Tony Crosland's plea that we should not scrap the Channel Tunnel project right away, despite the unfeasibility of the cost of the rail link, but await the Cairncross committee's report and allow him to negotiate a year's delay with the French'. The French Government was approached and M. Marcel Caraille, Secretary of State for Transport, wrote to Crosland on 9 December agreeing to the request for a revision of the timetable while emphasising that the French Government remained 'extremely attached' to the project. A bill authorising ratification of the Channel Tunnel treaty was in fact approved by the National Assembly on 17 December.

The fatal moment was approaching because under the terms of Agreement No 2, if the Anglo-French treaty was not ratified (which required the prior passage of the Channel Tunnel bill), the government was deemed to have abandoned the project. Some hectic correspondence took place. Crosland proposed to the French Government and the companies that the whole timetable should be put back to enable lower-cost rail options to be thoroughly examined: and that meanwhile ratification of the treaty should be delayed.

The companies felt, no doubt after taking legal advice, that they must protect their shareholders by serving the government with formal notice that the companies considered the project to have been abandoned and that the shareholders' rights to compensation were therefore enforceable. However, this formal letter was accompanied by another, explaining that, while the shareholders' rights had to be protected the companies were willing and even anxious to continue the project, and they suggested ways by which this could be arranged.

British and French civil servants examined this offer and – on the British side at any rate – objected to a proposal that shareholders could withdraw their money at a premium during the intermission period, and that those who remained could similarly withdraw at a premium if for any reason the project were abandoned before the main construction began.

The companies suggested a new timetable – a bill to be introduced in the autumn of 1975, to complete all its stages by the summer of 1976, with a commitment to start work by the end of 1976.

Having regard to the Labour Government's evident lack of enthusiasm for the tunnel, despite Anthony Crosland's personal commitment to it, the Channel Tunnel companies may be considered to have been justified in seeking protection for their shareholders in this way. But they had clearly indicated in private that their proposals were negotiable. However, the service of the formal notice gave the anti-tunnellers in the Cabinet the leverage required. On 16 January 1975 the decision was taken to accept the notice of abandonment and the liability to repay the shareholders their outlay on preliminary works plus a premium. Repercussions were of course felt in France where the government was unwillingly dragged into the same position vis-à-vis the French company. That point had been raised but brushed aside in Cabinet. *The Castle Diaries* relate that 'Typically, Tony would have preferred to keep the scheme ticking over without commitment but the reaction of the companies has made this impossible. So the crunch has come. Jim's [Callaghan's] was the only discordant voice. We were, he said, putting ourselves in a very difficult position with the French . . . he was worried about Tony's proposal to handle it "in this brusque way". . . . It was Tony who pointed out that the French had far more to gain from the project than we had; they did not face the need for a highly expensive rail link'. (It was curious that Crosland should apparently have forgotten the French promise to build their own high-speed rail link, Paris–Nord, which was much longer than British Rail's complementary link.)

So the fateful decision went through Cabinet with little argument; and the results of some twenty years of planning, negotiating, arguing, agreeing, raising money and spending it on major preliminary works, were thrown away, without regret except on the part of Crosland. On 20 January he told the House of Commons that 'the project will be run down as soon as possible. However, the studies, plans and works will be preserved in the best possible state so far as practicable in case the tunnel should be revived when circumstances are more propitious. Nothing will be done which might prejudice this possibility'. Appearing on television, he told viewers that he still hoped and believed that the tunnel would be built in his lifetime. Sadly, that was not to be; he died in 1977.

Mrs Castle recorded in her diary for 16 January 1975 'Personally I am relieved that Tony Crosland has decided we can't go ahead. This is not only anti-common market prejudice. It is a kind of earthy feeling that an island is an island and should not be violated. Certainly I am convinced that the building of a tunnel would do something profound to the national attitude – and certinly not for the better. There is too much facile access being built'. This revealing comment can perhaps be taken as an example of the curious reactions which the tunnel often seems to provoke – a former Minister of Transport who had sponsored the tunnel with her French counterpart, a future member of the European Parliament, expressing sentiments almost identical with those of Sir Garnet Wolseley and Admiral Sir William Horsey!

The abandonment of the tunnel illustrates the way in which public and political opinion tends to swing backwards and forwards, with at times pro-tunnellers and at other times anti-tunnellers in the ascendant. Harold Wilson, the Labour Party's leader, had announced a decision to go ahead in 1966; now returned to power after a Conservative administration had during four years supported the project, enthusiasm on his part seemed to have melted away. Crosland had certainly, after original scepticism, become converted, 'on the road to Damascus', as he once expressed it himself.

Justification for abandonment was claimed to exist on two grounds: the companies' attitudes to re-negotiation, and the need for British Rail to discover a cheaper rail link. Neither bears close examination. The companies had made it quite clear that their formal notice was merely a legal requirement, to protect their shareholders against the possible consequences of failing to serve such notice; they were willing to negotiate an entirely fresh agreement.

And on the rail link, British Rail had indicated clearly that, if the government so required, alternative and cheaper strategies could be put forward quickly. To suggest that British Rail had nailed its colours to the mast and demanded an independent, high-speed link as the price of co-operating in the tunnel project, would have been absurd. At one stage in the planning, no less than eight rail strategies, each with different levels of first cost, and different rates of return on a discounted cash flow basis, had been identified by BR's Channel Tunnel department. These were kept under wraps after the government had decided that the independent high-speed link should be built: but it was always recognised that a fall-back solution might be required at short notice. To blame BR, even by implication, for the need to abandon the project was quite unjustified, though several newspapers fell into that trap. The true cause must lie in the changing, uncertain world of politics where one year's enthusiasm becomes next year's scepticism, one year's energy next year's lassitude.

The Cairncross report appeared six months after the abandonment. It contained one telling sentence. 'We have been conscious as our work proceeded that everything seemed to be happening in the wrong order.' How true! At times, governments seem to act on the Alice in Wonderland principle: 'sentence first, verdict afterwards'. Certainly the Cairncross verdict on the tunnel was that it was viable, though a better integration of interests – between tunnel companies, governments and railways – was needed. This, British Rail planners could agree, was a glimpse of what to them had been long obvious.

15 Looking into a Mousehole

Throughout the long history of the Channel Tunnel, those who believed in the project never ceased to give up hope, even after such stunning blows as the Board of Trade's ultimatum to Sir Edward Watkin in June 1882 and the Labour Government's sudden announcement of abandonment in January 1975. Certainly the British and French project managers continued after 1975 quietly to study the possibility of reviving the project, in full consultation with BR and the SNCF. It was agreed that the best prospects lay in a proposal embodying:

(1) Minimum capital cost
(2) Minimum environment impact
(3) Minimum liability to opposition from competitors.

The indication was that a rail-only link in the form of a single-bore, single-line tunnel would meet these requirements. Technical studies by the railways were therefore put in hand. Just 3½ years after the débacle, BR and the SNCF had before them a report proposing a single-line tunnel linking their two systems, which was promptly nicknamed the 'mousehole'.

The report explained that, in the search for cost reduction, the engineers had initially considered using Southern Region third-rail electric traction in the tunnel, but it had emerged that the weight of trains, plus the speeds desired, in relation to aerodynamic resistance and the gradients to be surmounted, made this impracticable. Southern Region third rail moreover would require substations within the tunnel. The 25kV ac system used in other areas of BR and on the SNCF would not require substations for this length of track.

Consideration was given to using special tunnel locomotives, but this had major disadvantages. Utilisation would be poor, and a change of traction at each end of the tunnel would involve unacceptable lengthening of journey times.

Accordingly, in all the options considered, 25kV traction in the tunnel and through running from the English terminal to the SNCF was assumed. This indicated an increase in the tunnel diameter from the absolute minimum which the third rail might have allowed.

The next major question was whether a service tunnel in addition to the running tunnel was required; it had been specified in the abandoned 1975 project and it had three main advantages. It would speed up construction of the main tunnel; it would facilitate maintenance by reducing the length of time necessary to close the tunnel for that purpose in every 24 hours; and it would help passenger evacuation in the event of a breakdown. On the other hand it would of course increase the initial outlay.

Three solutions were therefore analysed: BR gauge without a service tunnel: UIC gauge without a service tunnel: and UIC gauge with a service tunnel. Allowing for fixed railway equipment and rolling stock, the capital costs were estimated at £518, £560 and £650 millions respectively. The environmental impact, such a sensitive factor in Britain, would be small in all three cases.

If BR gauge were used, an interchange station near Calais would be provided for night sleeping-car trains, but BR gauge stock hauled by dual-voltage locomotives would run through to Paris and Lille (for Brussels). Freight services would be, in effect, a replacement of those now provided by the train ferries, whereby wagons within the BR loading gauge run through to all Continental destinations. Single containers and small lots would however be transhipped at Calais between British and UIC gauge wagons.

On the other hand, the use of UIC gauge in the tunnel would facilitate its possible extension at some future date to London, which otherwise would be permanently ruled out. With UIC gauge in the tunnel only, BR trains and locomotives would work through to Paris and Lille (for Brussels) as under the first option; but the night sleeping car trains to and from the Continent would interchange with standard BR trains on this side of the tunnel, at Stanford (Westenhanger), six miles west of Folkestone. Freight would mainly be carried in wagons to the 'train ferry' gauge, but container traffic and trade cars would pass through the Tunnel in wagons to UIC gauge and be transhipped at Stanford.

The main effect on traffic of adopting the third option, namely the provision of a service tunnel, would be that trade cars with petrol in their tanks (and probably also Motorail traffic) could be accepted, because of the additional safety factor.

Opposition, which had been so vocal in 1973–74, should hardly reappear at all. The competing shipping services would continue their roll-on, roll-off business with relatively little diminution, though 'classic' or foot passengers using the ships would almost all divert to rail; they represented however a minor element in the profitability of the ships, compared with the roll-on, roll-off business. The modest terminal facilities required at Stanford would have only minimum impact, as would the contemplated use of either Victoria or Kensington Olympia for the tunnel passenger trains.

The tunnel, to railway operators, would in effect be a single-line section, 49.3 kilometres, (30¾ miles) long, in the main lines London–Paris and London–Brussels. Modern signalling with sophisticated speed controls can maximise the capacity of such a section. The problem of two-way reaction, ie the transference of delays from trains in one direction to those in the opposite direction – was however a real one. Breaking the section with a passing loop in the centre would not be of great assistance, taking into account the pattern of train services envisaged within the tunnel, though it would allow one or two late-running freight trains to be parked out of the way of the passenger trains. Ten minutes would be the timetable interval between the last train in one direction completing its journey and the first starting in the opposite direction, to allow for some late running.

The service through the 'mousehole' would be surprisingly adequate. It was planned to operate alternate 'flights', England–France and France–England, with a maximum interval between trains for any one service (London–Paris or London–Brussels) of 3 hours. Ten paths would be available in each flight lasting 45 minutes. Journey time would be 35 minutes; the tunnel would be closed for maintenance for six hours each night. At the opening, 40 passenger trains and 30 freight trains per day were planned, representing about 60 per cent of the tunnel's capacity.

The operating pattern included some special features. At any one time, a maximum of 10 trains would be permitted to be simultaneously in the tunnel. Every passenger train would be equipped with a driving cab at the end away from the locomotive, to permit complete reversibility. Freight trains would have locomotives both at front and rear, the rear locomotive being controlled by radio from the leading locomotive. In the event of a failure, reversing out of the tunnel could be achieved without delay. The overhead catenary, moreover, would be sectionalised into lengths of no more than two kilometres for isolation purposes.

The train service pattern was based on an off-peak period for about seven months of the year, a 'shoulder' peak of about three months and a high peak of about two months. Journey times between London and Paris were put at 4 hours 30 mintues and between London and Brussels at 4 hours and 10 minutes. On the basis of the fares proposed, it was calculated that about six million passengers would use the tunnel in the first year after opening and that subsequent growth would be between 2½% and 3% annually.

The proposed charges for freight traffic would offer a reduction below existing rates by road lorry and roll-on, roll-off shipping. On this basis an initial traffic of five million tons was assumed, with a subsequent annual growth rate of 3%. Trade cars – assuming a service tunnel to be provided – were put at 0.5 million tons in the first year.

The conclusion drawn from the estimates of capital cost, receipts, and outgoings was that the discounted rate of return on the 'mousehole' was quite attractive. Varying assumptions were worked into the calculations; the real rate of return varied, under these assumptions, only between 16.1% and 12.0%. Sensitivity tests were applied to measure the effects of variations in costs and receipts: the result showed that the project was robust in all the options.

By January 1981 the British Railways Board had committed itself to sponsoring the scheme. In a document entitled *Cross-Channel Rail Link,* circulated privately, it revised the traffic estimates and came down firmly in favour of UIC gauge in a tunnel with a diameter of around 6 metres (19ft 8in), together with a service tunnel of 4.5 metres (14ft 9in). Some BR/SNCF exchange facilties would be provided at Cheriton, with a passenger station at Saltwood. But the use of Kensington Olympia (in addition to Victoria) was now replaced by the proposed use of a site on the West London line at West Brompton, formerly a coal depot.

The required capacity on the Southern Region would be found chiefly by utilising the paths in the working timetable for boat trains, which would be replaced by tunnel trains. The line's capacity for tunnel trains was revised, as were the financial calculations and traffic estimates. The internal rate of return was shown now as 9.5% in the base case, falling to 7.0% if there was a shortfall of 30 per cent in total receipts.

Three months later BR went public with this proposal. The total cost was now put at £765 millions for the tunnel, spread over about seven years of the construction period and of course divided equally between Britain and France. The cost of BR infrastructure adjacent to the British tunnel portal would be about £83 million.

Thereafter for a considerable time the project hung fire. The Labour Government that had so abruptly cancelled the Tunnel, and which survived until 1979, was in no hurry to change its stance. The European Economic Community certainly displayed an interest, as it did in the case of all major transport links between member states. The EEC

Council of Ministers approved the setting up of consultative machinery to study the financing of major European transport links including the tunnel. But the new Conservative Government (of the same political complexion as the one that had signed the Anglo-French treaty and Agreements Nos 1 and 2) which returned to office in 1979, was committed to a monetarist philosophy which regarded public sector investment as inherently inflationary in its effects. The French Government was understood to be interested in financing its share of the costs of a 'mousehole' through the SNCF's investment budget, but such a solution did not appeal in the least to British Government thinking.

Slowly, however, and without departing from the doctrines of monetarist economics, the British Government began to view the tunnel as a possibly suitable candidate for private investment. Independent (ie, non-railway) financing for a 'mousehole' was for a time proposed by two groups in Britain, Channel Tunnel Developments (CTD) and the Cross-Channel Tunnel Group (ACTG).

But the major financial interests soon came to regard the omission of the vehicle ferry business as too seriously damaging to the financial attractiveness of the 'mousehole'. So, planning by various groups was resumed on lines that provided for road traffic, either by drive-through roadways or by means of ferry trains. The former option was generally reported to be favoured by the British Prime Minister, while the French attached more weight to the views of the SNCF than the British Government did to the position of BR.

At a meeting in London on 10–11 September 1981, both Mrs Thatcher and President Mitterand expressed enthusiasm for the concept of a fixed link, and agreed to the commissioning of a joint Anglo-French study of the options that were being put forward. (Compare this meeting with that of 8 July 1966 between Wilson and Pompidou, page 86.)

The report of the Anglo-French group was presented to the governments seven months later; it was published in Britain as a white paper. It was not a very conclusive document, though it agreed that an economic return could be expected from a fixed link based on bored tunnels (using proven technology), and that twin 7metre diameter tunnels were better than any 'mousehole'.

Soon afterwards (on 16 June 1982) the British and French Transport Ministers (David Howell and M. Charles Fiterman) announced that a decision would have to await the results of a study of the legal, financial and organisational aspects of a fixed link – as though all this ground had not been thoroughly explored in the 1960s and 1970s! (Compare with the Castle/Pisani meeting of 28 October 1966, page 98).

The House of Commons was told that any fixed link construction would, so far as the British Government was concerned, have to be undertaken without any Treasury financial guarantees and also – interestingly – that any traffic guarantees from BR would not be permitted since these might eventually fall back upon the government.

On the French side, no doubt bearing in mind the experience of *la perfide Albion* in 1975, M. Fiterman told the National Assembly that there must be a firm political commitment in Britain as well as in France if the scheme were to go ahead.

So the slow retracing of steps to the position of some sixteen years earlier was started. Once again, the formal minuet was to be danced. Promoters of various schemes would be encouraged to submit competitive proposals. There would be white papers and, probably, green papers; parliamentary committees would go over the ground yet once more.

History tends, certainly, to repeat itself but always with some differences. That was to be the experience of the next 4½ years.

16 Arriving at Eurotunnel

Between June 1982 and the autumn of 1986, there was great activity on the part of three major groups proposing a fixed Channel link, joined towards the end by a surprising new entrant, who had previously been an anti-tunnel campaigner.

It was interesting that several projects that had been more than once examined and discarded, re-appeared in a modernised form. The dream of building artificial islands in the Channel, first mooted by Mathieu-Favier in 1802, still seemed to tickle the fancy of some engineers. The desire to demonstrate the technique of immersed tubes rather than bored tunnels revived. The steel interests were enticed by the prospect of a huge bridge, as one option; the other option being the attempt to make the bridge more acceptable by combining it with a tunnel in the central area of the Channel.

Head and shoulders above the other projects, however, stood one firmly based on the research that had produced the former Channel Tunnel Study Group's report of 1960, and the immense amount of detailed work carried out since then in testing and refining the plan. It was compiled by a British consortium, CTG (The Channel Tunnel Group Ltd) with a French counterpart, France Manche SA. The group subsequently adopted 'Eurotunnel' as its operative title. It comprised a formidable list of backers. On the financial side, the National Westminster and Midland Banks participated, as shareholders, while merchant bank advisers comprised Morgan Grenfell & Co and Robert Fleming & Co. Five of the largest civil engineering companies in Britain were united on the construction side. (The full list of sponsors for all the projects submitted to the governments in 1985 is contained in Appendix A; so many names would clutter the text of this chapter.)

CTG/France Manche established the solidity of its backing by pointing out that the combined turnover of the British construction companies was more than £5,000 millions, and the assets of the two participating banks totalled £133,000 millions. On the French side, the backing was equally impressive, with the Crédit Lyonnais, the Banque Nationale de Paris and the Banque Indosuez heading five major contracting firms; the turnover and assets on the French side were even larger than on the British side. On the British side nearly 30 specialist consultants were retained to deal with individual aspects of the project.

The machinery of administration was planned to consist of a 'promoter' – a joint body of the two participating groups which would be the effective project manager; in law it would be an incorporated partnership (in France, *société en participation*). The two company boards and the partnership would consist of the same individuals, with the British and French chairmen being alternate chairmen of the partnership – thereby overcoming the problem created by the fact that a company can only be registered in one country of domicile. (Precedents of course exist in the great Anglo-Dutch companies, Unilever and Shell.)

In fundamentals, the project of Eurotunnel was that of 1960, as refined and developed during the early 1970s by RT-Z and the French project managers, but with one important difference. The British Government had now ruled out any substantial investment by British Rail in a high-speed link from London to the tunnel comparable with the French TGV network, and the estimates of rail traffic accordingly had to be based upon use of the existing Southern Region routes, with the limitations this would impose on speed and capacity.

Features of the scheme were a total length of tunnel of 49.2km, (30.7 miles) ie 9km (5.6 miles) between the British entrance and the shore line, 36.5km (22.8 miles) under the sea, 3.7km (2.3 miles) from the French shore to the French tunnel entrance. As in previous schemes, the configuration of the North Downs, and the road and rail access in Kent, led to the tunnel entrance being in the steep hillside and then leading down until the coast is passed below Shakespeare Cliff, west of Dover. (Earlier alignments had carried the tunnel further east, below Dover Harbour.)

The diameter of the running tunnels was slightly increased, to 7.3 metres (23ft 8in) internal diameter, with a service tunnel of 4.5 metres (14ft 6in) diameter. Cross adits would be provided every 375 metres, instead of 250 metres as in earlier schemes. The argument over the desirable number of rail crossovers to permit single line working had been finally settled: two crossovers were to be provided.

It was accepted that near the French coast some fissured ground would be encountered, and grouting by injecting a mixture of clay and cement into the fissures would be involved, as a preliminary to boring the tunnels. There was no final decision as to the use of cast iron or pre-cast concrete tunnel segements, the options being left open according to local conditions.

The British entrance in Castle Hill was now planned to be the main one, instead of, as previously, leading to a short tunnel through Castle Hill followed by an open stretch before entering the main tunnel at Holy Well. The tunnel was now to be continuous, with the shallow stretch between Castle Hill and Holy Well being constructed on 'cut and cover' principles. This would have environmental advantages and allow the Holy Well area ultimately to regain its tranquillity.

The Cheriton terminal was designed much on the lines of previous schemes; there would be 10 parallel lines and platforms, which road vehicles would reach from an overbridge; extension to 16 tracks at some future date would be possible. The procedure proposed was for vehicles entering from the M20 and A20 roads by means of new links to be segregated into separate streams for cars, coaches and lorries before passing, first, the toll booths and then the frontier controls for both Britain and France – (UK emigration, UK customs, duty-free shops, French immigration, French customs.) On arrival in France it would thus be possible to drive out directly on to the French road and motorway system. Similar arrangements would prevail in France, so that 'free exit' would exist in both directions.

Appropriate other amenities such as restaurants and cafés, toilets, motor repair and breakdown services, would be provided in the terminal area.

Loading the trains would be by driving through side doors in special loading wagons at the end of each rake of vehicles. The carrying vehicles would be of three types: double-deck car-carriers for vehicles less than 1.85 metres (6ft 2in) in height; single-deck vehicles for cars, coaches and caravans up to 4 metres (13ft 0in) in height, and lorry carriers for freight vehicles, also up to 4 metres high.

The car-carrying trains would generally comprise two rakes each of 13 car-carrying vehicles plus two loading-unloading wagons. Normally one single-deck and one double-deck rake would be coupled together; this would give a capacity of 126 cars in one rake plus 67 cars, or 13 coaches and 28 cars in the other.

The lorry-carrying shuttle trains would comprise 25 carrier wagons, plus the two loading-unloading wagons. Maximum lorry weights of 44 tonnes would be catered for.

Obviously questions arose about the problem of side-loading a shuttle-train, compared with the relatively simple end-loading practised on the ferry ships. It was arranged that bridging ramps from the access platforms would lead to sliding doors of ample width for even unskilled motorists to use.

Lorry shuttles would, in the words of a CTG leaflet, 'be loaded by

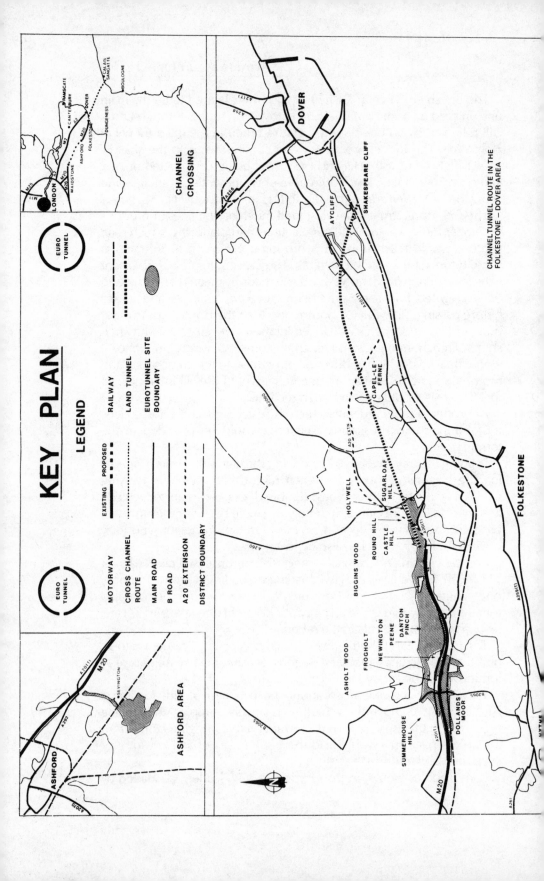

KEY PLAN

LEGEND

EXISTING   PROPOSED

MOTORWAY

CROSS CHANNEL ROUTE

MAIN ROAD

B ROAD

A20 EXTENSION

DISTRICT BOUNDARY

RAILWAY

LAND TUNNEL

EUROTUNNEL SITE BOUNDARY

EURO TUNNEL

CHANNEL CROSSING

ASHFORD AREA

ASHFORD

M20

SEVINGTON

A20(T)

A292

A2070

DOVER

A2171

A256

A258

SHAKESPEARE CLIFF

AYCLIFF

A2011

CAPEL-LE-FERNE

A20 EXTN

B2060

HOLYWELL

SUGARLOAF HILL

A20(T)

ROUND HILL

CASTLE HILL

BIGGINS WOOD

A260

ASHOLT WOOD

FROGHOLT

NEWINGTON

PEENE

DANTON PINCH

A20(T)

A259(T)

B2065

DOLLANDS MOOR

B2065

SUMMERHOUSE HILL

A20(T)

M20

A261

FOLKESTONE

HYTHE

CHANNEL TUNNEL ROUTE IN THE FOLKESTONE – DOVER AREA

CHANNEL CROSSING

LONDON

M25

M11

M2

M20

MAIDSTONE

ASHFORD

FOLKESTONE

DOVER

CANTERBURY

RAMSGATE

CALAIS SANGATTE

BOULOGNE

DUNGENESS

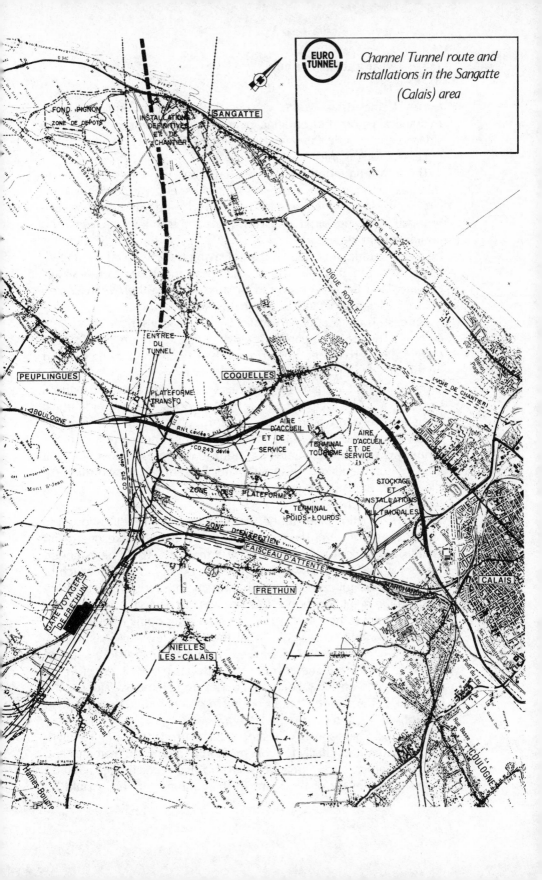

EURO TUNNEL

Channel Tunnel route and installations in the Sangatte (Calais) area

lorries being driven onto a special flat wagon equipped with hydraulically-operated bridging ramps and stabilising jacks on both sides to enable articulated vehicles to utilise the opposite platforms as necessary for loading'.

The transit time between the two terminals was expected to be 30 minutes; although the maximum running speed would be 160km/h (100mph), slow running would be involved round the loops at each terminal. Allowing for passing toll booths and frontier formalities, and awaiting train departures, an average elapsed time of 55 minutes between entering one terminal and leaving the other was estimated, apart from the possible use of duty-free and other non-essential facilities, and passengers would be able to leave their vehicles and utilise facilities provided in the trains such as toilets and vending machines. Train attendants would look after passenger safety and comfort. The time compares with the average 2½ hours of 1986 from entering the ferry port on one side of the Channel to leaving the port on the far side using a car ferry ship.

Using 18 shuttle train sets, the design provided for a throughput capacity of 1000 cars per hour at peak periods. Bringing additional trains into service would enable the throughput to be considerably expanded in line with rising traffic demands. The signalling in the tunnel would enable 20 trains per hour to run in each direction, ie a three-minute headway. This could later be reduced to two minutes, allowing 30 trains per hour but, safely, three miles or more apart at 160km/h (100mph).

The construction of the tunnel would follow, with modifications, the plan envisaged in 1974. The existing site at the foot of Shakespeare Cliff established by contractors for RT-ZDE at that time would be extended and developed, and railway sidings would be laid to enable much of the materials required to be brought by train, thus reducing heavy lorry traffic in the neighbourhood. This site was connected in 1973–74 with the main Folkestone–Dover road by means of a sloping road tunnel.

Other working sites would be at the upper level of Shakespeare Cliff, at Cheriton, and at Holy Well, where the main bored tunnel would follow the cut-and-cover section. Because of differences in the geography and the length of tunnel under land on the two sides of the Channel, it was planned that six boring machines would be at work on the British side, and five on the French.

The upper Shakespeare Cliff site would be used for sinking a shaft to the main tunnels, initially for construction workers and materials and later as an emergency ventilating shaft. Normal ventilation, in addition to the piston effect of the trains in the running tunnels, would be

FREIGHT and PASSENGER TRAIN ROUTES
INFRASTRUCTURE and TRAIN MOVEMENTS
Waterloo to Brixton and
Willesden to Brixton and Falcon Junction

		DAILY TRAINS EACH WAY			
		Present		with Tunnel	
		Winter Weekday	Summer Saturday	Winter Weekday	Summer Saturday
▬ Passenger Route Victoria via Brixton	SR Domestic	215	173	215	173
	International	9	19	0	0
⋯ Passenger Route Waterloo via Brixton		0	0	30	44
▬ Freight Route Willesden via Brixton	SR Domestic	12	6	12	6*
	International	2	2	12	12
▬ Freight Route Willesden via Falcon Junction	SR Domestic	5	5	5	5
	International	0	0	9	9
▨ Passenger train empty stock route	International	0	0	15	22
▬ ▬ New international passenger train chords					

⋯⋯ Alternative international passenger train route

▬ ▬ Not used by international Channel Tunnel trains

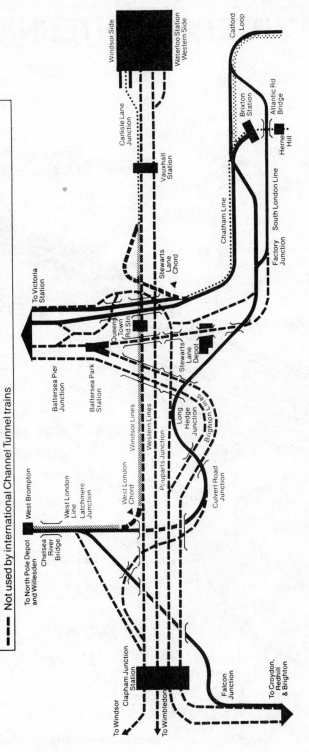

To North Pole Depot and Willesden
West Brompton
Chelsea River Bridge
West London Line
Latchmere Junction
West London Chord
To Windsor
Clapham Junction Station
To Wimbledon
Falcon Junction
To Croydon, Redhill & Brighton
Pouparts Junction
Culvert Road Junction
Long Hedge Junction
Brighton Lines
Battersea Park Station
Battersea Pier Junction
Windsor Lines
Western Lines
To Victoria Station
Queens Town Rd Stn
Stewarts Lane Depot
Stewarts Lane Chord
Vauxhall Station
Chatham Line
Carlisle Lane Junction
Windsor Side
Waterloo Station Western Side
Catford Loop
Brixton Station
Atlantic Rd Bridge
Herne Hill
Factory Junction
South London Line

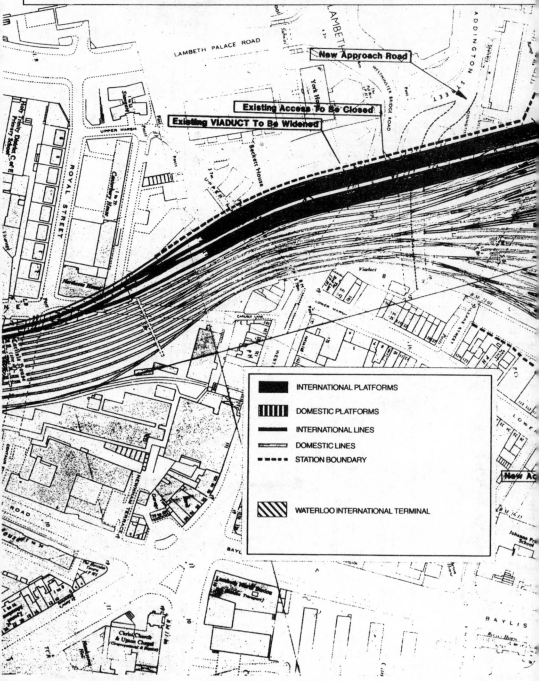

Plan of the proposed Waterloo International Station (British Rail)

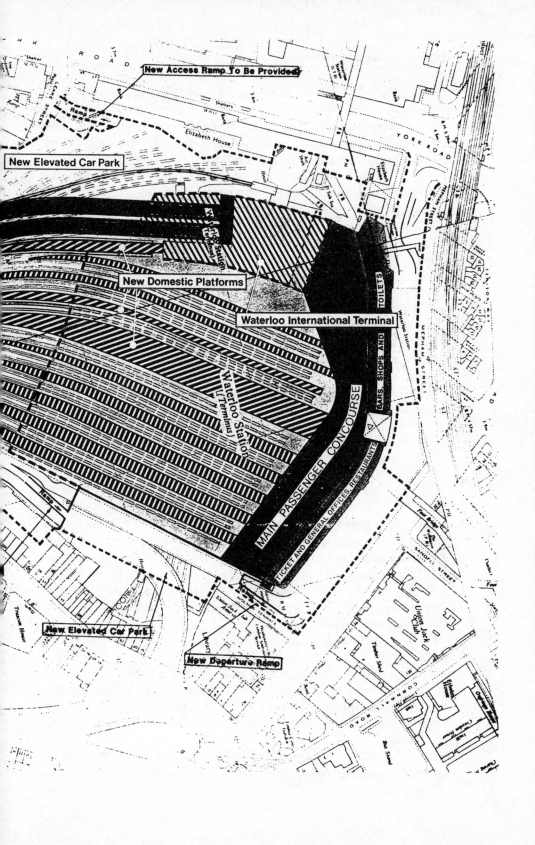

New Access Ramp To Be Provided

New Elevated Car Park

New Domestic Platforms

Waterloo International Terminal

Waterloo Station (Terminus)

TOILETS

BARS, SHOPS AND

MAIN PASSENGER CONCOURSE

TICKET AND GENERAL OFFICES, RESTAURANTS

New Elevated Car Park

New Departure Ramp

Elizabeth House

YORK ROAD

MEPHAM STREET

Waterloo Station

SANDELL STREET

Union Jack Club

Library

provided by input of air to the service tunnel, creating a plenum, with exhaust through the running tunnels to the portals. Construction of the Cheriton terminal would involve filling the site of 140 hectares (260 acres) with some 500,000 to 700,000 cubic metres of spoil. Some environmental advantage would result from this being obtained from the spoil heaps of collieries in the East Kent coalfield. In addition to Cheriton, a road freight inland customs clearance depot is planned for Sevington, south east of Ashford, in an industrial development zone.

On the French side, a terminal similar in principle to Cheriton would be provided at Fréthun, although the less congested nature of that area permits a more open arrangement, particularly of the rail connections to both the Lille and Boulogne main lines of the SNCF. The French entrance to the tunnel, moreover, instead of being in the form of an adit entering a hillside, is a *descenderie,* the line falling in cutting until it enters the tunnel well below the level of the surrounding area. construction, a |shaft ' will be sunk from the ground surface near Sangatte to tunnel level.

The total investment (half in Britain) would be £5.4 billion which would include £1 billion for unforeseen contingencies. This total was estimated to cover:

(a) capital construction costs, allowing for inflation;
(b) financing costs during the construction period;
(c) interest during the construction and initial operating period;
(d) contingencies.

The equity element in this was expected to reach £1 billion. If it fell short of this amount, the debt would be correspondingly increased. But debt and equity were inter-related, in that the equity would not be provided unless there was a guarantee of loans adequate to see the product through to completion; and the loans would not be forthcoming without a minimum level of equity capital. Letters from the participating banks, merchant banks and brokers stating that the necessary finance has been committed or promised had been obtained.

There was also a firm financing plan for raising the equity in three 'tranches', the first as 'seed money' to be underwritten by the members of Eurotunnel; the second by a private placement; and the third by public subscription, enabling the main construction work to be started.

How would the investment by remunerated?   Promoters' consultants forecast total cross-Channel traffic as follows:

**Cross-Channel passenger traffic in millions,
1993 (opening year of Tunnel)**

	Total	Tunnel	Tunnel Share %
Car passengers	9.9	6.3	63.6
Coach passengers	8.8	4.4	50.0
Day trip parties	3.9	3.1	79.5
Other (including air & through rail to W. Europe)	47.9	15.9	33.2
TOTAL	70.5	29.7	42.1

Cross-Channel freight traffic in million tonnes, 1993

	Total	Tunnel	Tunnel %
Ro-Ro freight	24.2	6.0	24.8
Containers and rail wagons	7.9	4.0	50.6
Bulk (including trade cars)	41.8	3.2	7.7
TOTAL	73.9	13.2	17.9

At 1985 prices, the tunnel gross receipts were expected to be:

	£ million
Passengers	288.8
Freight	107.3
Ancillary	35.7
TOTAL	431.8

About one-half of the ancillary receipts were expected to come from duty-free sales.

Operating costs were estimated at £74 million in the first year after opening.

Average charges for tunnel transit in 1993 were calculated – on a basis broadly comparable with those charged by competing services – to be as follows, at 1985 prices:

	£
Car passengers	19.80*
Coach passengers	5.70
Excursionists by coach	3.50
Through rail passengers	8.00
Freight	
Ro-Ro, per gross tonne	10.00
Container/rail wagon per gross tonne	9.20
Trade vehicles/rail wagon per gross tonne	8.80
Bulk train/rail wagon per gross tonne	1.40

*including an element for the vehicle.

The British Railways component in the project, covering infrastructure, rolling stock, services and receipts is dealt with in Chapter 17. It is important however to note that so far as Eurotunnel is concerned, the shuttle trains are planned to be built and operated by the tunnel, not by the railway administrations; the tunnel receipts from through rail passengers and freight will comprise tolls paid by the railways to the tunnel, and recovered by the railways through their own fares and freight rates.

The CTG/France Manche timetable for construction was, assuming that the necessary legislation was passed by March 1987:

Construction begun	Mid-1987
Service tunnel complete	Spring 1990
Running tunnels complete	Spring 1991
Start of commissioning	Autumn 1992
Public service opens	Spring 1993

The Eurotunnel project was accompanied by a very detailed report, *Environmental Effects in the UK* dated October 1985. The promoters had appreciated the part played by environmental issues in the decision to cancel the earlier project, and were determined to defuse the opposition as far as possible. Apart from this, the government's official invitation to the promoters of all fixed link schemes had required an environmental impact assessment (known as an EIA) to be completed.

A firm of consultants, Environmental Resources Ltd, undertook the preparation of the EIA which was a massive document, covering every aspect of the subject. Only a few instances can be given: Kent County Council had a preference for the tunnel spoil to be used to fill up disused aggregate workings at Dungeness but this was objected to by ecologists. A preferred option was the strengthening and extension of the foreshore, now subject to erosion by the sea, at Shakespeare Cliff. Induced development in the neighbourhood of Cheriton was studied, as was the effect of reduced traffic flows within Dover. The effects on the population of the hamlets of Newington, Peene and Frogholt (total, 195) was analysed, and also the effects upon agricultural holdings – five at Cheriton, and one each at Saltwood, Ashford and Shakespeare Cliff. It was shown that 124 hectares (about 250 acres) of land would be permanently lost, of which only 55 hectares would be Grade 3A or higher agricultural quality. Screening by earth banks and plantings to minimise noise and visual intrusion were outlined. In all, 18 separate studies were undertaken.

The other competing projects all proposed drive-through facilities for road traffic. They had a number of interesting features. 'Euro Route' was a combination of two groups, Euro Route Ltd in the United Kingdom and Euro Route France on the other side of the Channel. The 'concession companies' participating were quite impressive; on the British side they included Barclays Bank and the financial conglomerate Trafalgar House, with interest also being shown by British Steel and Associated British Ports. In France the Banque Paribas and the Société Générale were among powerful financial interests, with Usinor representing steel, and Alsthom construction technology. Two construction consortia backed up the concession companies – Euro Route Construction Ltd and Scoltram SA in France.

Euro Route had sought to overcome the objections – previously fatal – to a bridge by reviving the concept of two bridges from the two coasts, 8.4km and 7km long respectively, reaching artificial islands which would be joined by an immersed tube tunnel some 21km long, thus leaving an open seaway in the centre. On each island the descent to the tunnel would be on a gradient of less than 4% (1 in 25) in the form of a spiral roadway, 2km long. The two main shipping lanes would be divided by a central ventilating shaft rising from the sea floor.

It was calculated that a mixed flow of cars and commercial vehicles could pass through the bridges and tunnel at a maximum rate of 3000 per hour, though speed restrictions would be enforced: 100km/h (62mph) on the bridges, 50km/h (30mph) on the spiral ramps, and 80km/h (50mph) in the tunnels.

The problem of high winds on lofty bridges would be dealt with by installing wind deflectors and if necessary imposing restrictions on empty high-sided vehicles. Access to the bridges in England would be from the M20 motorway and the A20 trunk road. An inland customs clearance depot for heavy lorries would be sited near Ashford.

In addition to the road bridge-cum-tunnel, a 'mousehole' rail tunnel was proposed, with a second rail tunnel to provide two tracks being built later.

The capital cost of this scheme was put at £3.7 billion for the motorway and £1.5 billion for the rail tunnel. Inflation, interest charges during construction and an allowance for cost over-run would bring the total to £9.7 billion. It was clearly designed to appeal to those politicians and industrial interests who favoured drive-through facilties for motor traffic; apart from that it appeared to be doing in a complicated and expensive way something that could be done more simply and cheaply.

The same object inspired the promoters of 'Eurobridge', a project that ignored the heavy and previously fatal objections to a continuous bridge across the Channel – the interference with shipping, especially during the construction period, the need to secure widespread international agreements, the hazards created by wind and fog and the terrible consequences of a monster tanker colliding with a bridge pier. It was however largely backed by two giants of the chemical industry, Imperial Chemical Industries and Dupont Fibres, since it was desired to use the bridge to demonstrate the properties of a new cable fibre which these firms had developed. It was proposed to build a huge suspension bridge providing twelve lanes for road traffic – though one lane might be used for a single-track rail link if the governments so wished. To overcome the problem of side winds affecting motor traffic on the bridge, the motorway would be enclosed. Drivers would pass through, in effect, a sort of aerial tunnel. Each main bridge span would be 4.5km (2¾ miles) long, using 'parafil' suspension cables – three times as long as the longest suspension bridge at present open.

The total cost was put as between £5.0 billion and £5.9 billion at 1985 prices. Allowing for inflation and interest charges during construction, the total would be about £9.0 billion.

Last on the scene came the surprising 'Channel Expressway', sponsored by the Sea Containers group based in Bermuda, which had purchased Sealink UK from British Rail under the Conservative Government's privatisation policy and which had, through Sealink, vigorously opposed any form of fixed link.

Its main feature was bored tunnels, 11.3 metres (36ft 8in) wide which would provide an underground motorway and, originally, space for railway trains running on tracks embedded in the hard shoulders of the motorway. Not surprisingly, both BR and the SNCF had rejected this notion and 'Channel Expressway' was remodelled to provide a separate rail tunnel.

The sponsors comprised Sea Containers on the British side and SCREC – a consortium of construction firms – on the French side, with financial support from Crédit du Nord. The cost of the project was somewhat optimisitically put at £2.55 billion.

It raised a number of difficult questions. First of all, Sealink had been very active in promoting the so-called 'Flexilink' campaign against a fixed link. Was this a kind of death-bed repentance? Alternatively, was the project merely thrown into the ring to make the task of selection more difficult, a diversionary tactic on the part of the shipping interests?

The French Government was reported to be sceptical about the extent of any genuine French involvement in financing this scheme;

both British and French experts considered the cost estimates to be unrealistically low.

The British Government requested the all-party transport committee of the House of Commons to report upon these four projects before a choice was made. In November 1985 the committee considered written memoranda from the promoters and took oral evidence from them. But first of all, evidence was taken from representatives of 'Flexilink', the pressure group opposed not only to a tunnel but to any fixed link. The memorandum from the group suggested that 'predatory pricing', 'excess capacity' and 'de-stabilisation of the market' would occur. It was argued that the ability of the sea operators to maintain services outside the range of the tunnel's competitiveness – for instance from Portsmouth to Brittany – would suffer from the loss of the profitable business on the Dover Strait. The Dover Harbour Board stressed the loss of employment in Dover. The sea operators also argued that their intention of building new, larger ships would enable them to offer better services. (It was pointed out by the committee that presumably larger ships would sail at less frequent intervals away from the periods of peak demand, so that waiting time would be much more than with the tunnel.)

The submitted schemes were then discussed with the promoters. The Eurotunnel team was led by Sir Nicholas Henderson, chairman of CTG. The committee took the memoranda as read, but particular attention was paid to environmental questions including so remote an issue as the possible entry of rabid dogs or foxes through the tunnel. The Eurotunnel group provided particulars of control measures such as the exclusion of animals from the terminal areas with the use of electrified mesh, the prevention of any sewage from train lavatories, or discarded food scraps, being deposited in the tunnels, and the regular patrolling of the tunnels within the maintenance procedures.

The project which aroused most scepticism in the committee was 'Channel Expressway'. The witnesses' answers to questions revealed that *no* financing had yet been arranged (Question 113); that the rail link had only been included because it was felt that the French would insist upon it (Question 117); and that no construction company had committed itself to the scheme (Question 114), merely that SPEA, an Italian firm with experience of tunnelling had the confidence of the promoters (Questions 129).

Ventilation was the technical issue most closely pressed and here the promoters replied that it was intended to use electro-static precipitators – a new technique which had been used in Japan for a tunnel as long as 12km (7¾ miles). The committee doubted the validity of the cost estimates.

The committee's report was published on 5 December 1985. Not surprisingly, it recommended that a number of measures should be taken to protect the public interest, particularly in environmental matters. It recorded that Kent County Council had decided that the CTG scheme appeared to be 'the least damaging to Kent's interests, so far as the East Kent economy is concerned'. On the economic case, the committee felt that, unless the government regarded the provision of a drive-through road link as indispensable, the CTG/France Manche scheme should be approved. But Channel Expressway was dismissed with the words that it 'presents a greater challenge to credibility . . . the cost estimates . . . are low in comparison with the other schemes without any explanation'.

The committee did not accept the Flexilink arguments; the tunnel was a robust project and 'because the link cannot be driven from the market even by bankruptcy, the costs to the ferry operators of fighting against it would be insupportable'. However, 'the government should ensure that a mandate to proceed with the scheme provides safeguards against the predatory pricing and abuse of an effective monopoly position'.

The committee also decided that it could not recommend the Eurobridge project. Its report went on to say: 'That leaves only the proposals of Euro Route and Channel Tunnel Group. While attracted by the drive-through facility, the committee is somewhat concerned about the financial viability of Euro Route and whether the user would in fact derive any financial benefit from its fixed link. On balance therefore the committee's preference is for the proposal of the CTG.'

The committee recommended that legislation should follow 'hybrid bill' procedure and that a time-consuming public enquiry was unnecessary, as petitions against the bill would be heard, as in the case of private bills, by a select committee. On 9 December the House voted by 227 votes to 181 approving the report.

Finally, at a joint ceremony in Lille on 20 January 1986 President Mitterand and Mrs Thatcher announced their support for the project put forward by the CTG/France Manche consortium. (Compare the Peyton/Chamant announcement of March 1971, page 101.)

Mrs Thatcher described the event as 'historic' and 'a dramatic step in Anglo-French co-operation'. She added that the CTG/France Manche scheme had been the best-researched and the most detailed plan, and 'had the best chance of winning the necessary financial support'. However, her agreement had apparently been purchased at a price: the Group promised to present proposals for a drive-through road tunnel before the year 2000. Its concession would last only until 2020.

Mrs Thatcher was thus able to maintain the position that a drive-through road tunnel was the ultimate aim.

Soon afterwards, in Canterbury this time, an Anglo-French treaty was signed. Meanwhile the French Government had announced on 22 January 1986 a regional development programme 'to mobilise the vital forces of Nord/Pas-de-Calais, Picardie and Haute Normandie', including connection of the tunnel with the motorway network. A British Government white paper appeared on 4 February describing the expected effects of the tunnel upon employment and promising completion of the M20 motorway link, and was approved in the House on 10 February by 268 votes to 107. Five Conservative members (four from Kent constituencies) voted against the government and about 60 Labour members abstained.

On 14 April 1986 the Channel Tunnel Bill (48 clauses, 6 schedules) was introduced in the House of Commons, under the hybrid bill procedure. Its passage would be the key to unlocking the door and enabling contracts to be let for the preliminary construction works and in addition – most important – it would enable the Anglo-French treaty to be ratified and take effect.

What had been the net effect of this long-drawn-out process? In essentials, it was a return to the CTSG proposal of 1960 which, given a more consistent and courageous government attitude in Britain, could have seen the tunnel in operation well before 1970, and built at a cost far below that which now must be faced. Private financing had now re-appeared, and the unnecessary complication of the nationalised Channel Tunnel Operating Authority disappeared. In anticipation of the passage of the Channel Tunnel Bill in Britain, and the complementary (though far simpler) measure in the French National Assembly, the British and French contractors joined in a planning consortium entitled (a nice example of 'Franglais') 'Transmanche Link'.

So far as the design of the tunnel was concerned, only detailed improvements of no fundamental importance had been made, though especially refining the arrangements for ventilation, spoil disposal and environmental protection.

The rail planning also represented in some respects a return to the ideas of 1960, but here some significant improvements, even though falling short of a British Ligne à Grande Vitesse on French principles, had been incorporated. Nevertheless planning of 'Rails into Europe' began to take on a more positive look as described in the following chapter.

17 Rails into Europe – at Last

Before the second world war, the railways and their associated shipping fleets enjoyed, a leading position, if not a near-monopoly, on the short-sea routes between England and the Continent. After the war this market expanded enormously, and also changed, year after year. Several factors were involved.

First, the time-conscious business traveller virtually ceased to use surface transport and transferred to the airlines. Next, the huge expansion in package holidays organised by travel agents was largely, though by no means exclusively, based upon low-cost charter air travel. The other great expansion in the holiday trade was Continental motoring, using the car-ferry shipping services. Lastly, the expansion in European international trade was chiefly catered for by heavy goods vehicles, also using the ferry ships.

Even though the railways carried a much diminished share of the total traffic passing, there remained a solid core of 'classic' passenger travel by rail and sea. It was partly based on cost (rail-sea fares being lower than air fares) and partly to a dislike of air travel among a proportion of travellers. It survived, surprisingly, the deterioration in standards caused by the concentration of the shipping services upon RoRo traffic and the reduction in special boat trains connecting with the ships.

On the freight side, though BR failed to share in the total market growth, the train-ferry wagon services survived, with some 10,000 wagons in the total pool owned by British Rail, European railways and such specialist agencies as Transfesa and Interfrigo. The large Transfesa depot at Paddock Wood in Kent, for instance, continues to handle large quantities of imported foodstuffs from Southern Europe in refrigerated wagons. About two million tons of Continental freight passes annually over British Rail.

Formerly, BR's shipping subsidiary, Sealink UK, contributed profits which to some extent offset the failure of the railways' international traffic to share the post-war growth. The enforced sale of Sealink UK under the government's privatisation policy made it even more desirable to exploit the possibilities of a fixed link.

Both on the passenger and on the freight side, BR is well placed to attract a very substantial traffic back to rail once the inconveniences of inter-modal transfer – train to ship and vice versa – are eliminated and journey times are reduced.

The strategy worked out by BR, the SNCF and Eurotunnel is a compromise between (a) the minimum investment proposals of 1960, which envisaged a 'frontier point' between BR and the SNCF at Westenhanger with 'all change' between trains for passengers; and (b) the full-blooded Ligne à Grande Vitesse of 1973–4, which in effect would have brought Continental trains to London. It also differs in several ways from the 'mousehole' proposal of 1981; it may be considered as an ingenious middle-of-the-road strategy designed to yield the maximum traffic increment, within financing limits that produce a true commercial return to BR and are thus acceptable to the British Government.

Central to the scheme is the agreement of the SNCF and other Continental administrations to accept passenger rolling stock and locomotives to the British loading gauge – though of course with couplings, buffing gear, brakes, etc conforming to UIC standards. Designing such stock presents no problems – the Night Ferry ran, apart from the war period, for almost 35 years between London and Paris. But a new development is that poly-current locomotives capable of operating on the Southern Region, the SNCF and SNCB will be built, which obviates a locomotive changing stop at the tunnel terminal. Triple-voltage locomotives – 750V dc third rail, 25kV ac overhead, and 3,000V dc overhead – will allow through working without change of traction to Brussels as well as Paris to take place. It is in fact technically possible to build locomotives that can operate on five types of current, adding $16\frac{2}{3}$Hz ac and 1,500V dc to the three systems already listed, which would enable through running to Cologne over the Deutsche Bundesbahn or over the Netherlands Railways to be established if commercially and operationally desirable.

This change of heart on the part of the Continental administrations can be attributed to a realisation that the British Government is adamant in rejecting the investment required to provide a separate route from the tunnel to London, built to the Continental loading gauge (see Appendix B). It is nevertheless noteworthy that the £373 million estimate, which caused near-apoplexy in ministerial circles in 1974, has been replaced by a total BR investment requirement, for to-day's more modest proposals, of £400 million at 1986 prices.

The present strategy has discarded both Victoria and White City as possible London terminal sites. Instead, Waterloo which was considered and rejected in 1970 has now been adopted. An important contributory factor has been some reduction in traffic and frequency of service over the Windsor lines serving the Hounslow loop, Richmond, Windsor and Reading. This has helped to make it practicable to re-model platforms 16–21 to provide a new Waterloo International terminal. The suburban trains displaced will be handled at platforms 14 and 15 and at two new platforms to replace the wide cab road between platforms 11 and 12. This spacious area is no longer used for taxis meeting arriving trains, its original purpose when built, and the Post Office and parcels vans will be handled elsewhere. (Railway historians will recall that this was the site, in the old 19th century Waterloo, of the through line to the former SER Waterloo Junction.)

The long new international platforms needed for the tunnel trains will be extended at the country end, and the buffer-stops moved forward, to provide space for a reception area and departure lounge, with all the facilities associated with a major airport terminal.

In finalising the design, much depends on the attitude of HM Customs and Immigration. If they can be persuaded to carry out frontier formalities on the trains in motion, as is the normal practice in Europe to-day (in some EEC countries frontier passenger controls have virtually disappeared), then controls at Waterloo will be unnecessary, apart from some accommodation for the very exceptional case where a passenger might be detained on arrival for questioning. Failing this solution, the former Night Ferry practice of passing passengers through frontier controls at the London terminal will presumably be followed. The trains will have sealed windows with air conditioning and power doors under the control of the train staff, so that full security between the tunnel and London would be maintained. Interview rooms on the trains can be provided for the use of customs and immigration officers who might need to interrogate any passenger in private.

From Waterloo the Windsor line tracks will be shared with local trains but there is some space for additional tracks on the site of the vanished goods line into the former Nine Elms depot, now replaced by the new Covent Garden Market. At Stewart's Lane a new flyover will connect with the Chatham main line which has benefitted from new connections that enable the parallel South London line tracks, under-utilised in comparison, to be used in common with those of the Chatham line between Stewart's Lane, Brixton and Peckham Rye. From Brixton effectively four tracks (considering the Catford Loop as part of the main line) are available through Shortlands to the Bickley junctions, where the former South Eastern main line is followed to Tonbridge on the No 1 boat train route.

From Tonbridge the route runs almost dead straight to Ashford, never deviating by more than half a mile from a line ruled on the map between those points. It was laid out by Sir William Cubitt for the former South Eastern Railway to be ideal for fast running, with platform loops available to allow expresses to overtake slower trains at several stations. (Ironically, the South Eastern had an overall speed limit of 60mph, even on this splendid racing ground.) Tunnel trains will traverse it at 100mph; higher speeds would be possible so far as the track geometry is concerned, but confliction with slower Southern Region services might present problems.

At Ashford an ingenious solution has been found. There are through tracks in the middle of the station, now used mainly by freight trains and Continental boat trains; otherwise Southern Region internal services use the platform loop lines. The centre lines will be carried beyond the existing station to serve a new Ashford International station which will be an island platform 400 metres long; the slow lines from the existing station will pass outside the International complex.

Ashford International will cater for the quite substantial number of travellers from the South East who would not wish to go to London to join tunnel trains. It will also provide interchange with Southern Region internal services. Through tunnel trains from and to the provinces will call here. Substantial car parking facilities, short stay and long stay, will be available.

The tunnel trains will diverge from the Ashford–Folkestone line at a new junction at Saltwood, from which they will head straight for the Cheriton complex and the tunnel entrance. Freight sidings are to be built at a point known as Dolland's Moor, at a location between the east portal of Saltwood Tunnel and the new Saltwood Junction. It also lies between the existing railway and the M20. They will be used for holding trains to and from the tunnel. Locomotive changes, for other than the non-stop tunnel passenger trains, will take place here, as will the customs inspection of seals on wagons and containers consigned to inland clearance depots, together with safety checks on inward and outward bound freight services.

Passenger services between the Midlands and North and the Continent are planned to run via the West London Line which connects with the various Southern Region main lines in the Clapham Junction area. The West London will be electrified on the Southern Region 750V dc third rail system and also the 25kV ac overhead system, the changeover point being Kensington (Olympia).

The through provincial trains will stop in Kensington Olympia, a station which has already taken a new lease of life from the north-south long distance trains which serve it. It will need further improvement for Channel Tunnel traffic.

Servicing and maintenance of the tunnel trains is planned to be carried out at a site on the West London line near North Pole Junction (Old Oak Common); there is still in situ a disused railway embankment linking the West London line with the Windsor lines into Waterloo (a relic of the old Waterloo–Richmond service via Addison Road, Hammersmith and Gunnersbury, withdrawn in 1916) which can easily be reinstated for empty carriage stock movements into and out of Waterloo.

At the north end of the West London line, Willesden freight yard (adjoining the Freightliner terminal) will be enlarged so that trains to and from the tunnel can be marshalled, and customs clearance facilities provided for traffic not consigned to inland clearance depots.

Construction of the traction and rolling stock clearly presents a challenge; British Rail Engineering's workshops will be in competition with private builders in this country and also with Continental manufacturers. The design of poly-current locomotives is well known on the Continent but is less familiar here, though dual-voltage operation at 750V dc third rail and 25kV ac overhead on the Great Northern suburban services into Moorgate has been employed for some years, and a similar combination is to be used on the revived cross-London services between the SR and the Bedford line via the Widened Lines.

The passenger rolling stock will have to be built to the limitations in dimensions which are discussed in Appendix B. The straight stretch of railway between the tunnel and Tonbridge (and probably also Redhill) could be modified to accept UIC 'X' and 'Y' stock probably at relatively modest cost: but unfortunately enlarging the loading gauge from Tonbridge to London would involve unacceptable problems and outlays, certainly in the forseeable future. With a larger than Berne gauge Channel Tunnel, enlarging the clearances for UIC stock to London could only be considered in later years should there be commercial justification.

The train service contemplated is a basic hourly interval between London and Paris in 3 hours 15 minutes, and London and Brussels in 2 hours 55 minutes, supplemented by up to four trains per hour each way at times of peak demand. About half the pathways through the tunnel will be reserved for international train services.

The Channel Tunnel will more than halve today's classic times for foot passengers and increase frequency; in the summer of 1986 there were just four London–Paris train/boat/train services by the short sea route, two via Folkestone–Boulogne and two via Dover–Calais and one of those was a car-ferry service from Dover Eastern Docks with bus connection from Dover Priory, the fastest taking 6 hours 58 minutes. In addition were four hovercraft services via Dover–Boulogne, the fastest taking 5 hours 15 minutes between London Charing Cross and Paris, but subject to reasonable weather. Between London and Brussels the cut in time will be much more by comparison with the 1986 7 hours 28 minutes by train/ship/train and 5 hours 10 minutes by the Dover–Ostend jet hyrofoil link.

The new times via the tunnel assume construction by French Railways of its long projected high-speed route from Paris to Lille and the Channel Tunnel. Had the tunnel not been cancelled by the British Government in 1975, that route – the first section of a new international high-speed link between Paris, Brussels, Amsterdam and Cologne – would have been built as the SNCF's first TGV route. The events of 1975 induced the SNCF to build first the Paris–Lyon TGV (Paris–Sud-Est) and then the line to Brittany and the South West, (TGV Atlantique). But the intention to build the Paris–Nord TGV always remained; it awaited a decision on the tunnel.

The London–Paris and London–Brussels timings should be competitive with air and attract a good proportion of daytime business travellers. Through services from provincial centres will depend upon the willingness of HM Customs and Immigration to carry out on-train examination. But night sleeper trains from Waterloo to Germany and Switzerland are planned, with at least two other trains at night which will stop at Fréthun on the French side of the tunnel, where interchange will be made with standard Continental sleeping-car and couchette trains to Southern France, Spain and Italy.

Motorail services to Continental destinations may start from Ashford or Cheriton; depending on demand, other starting points may be identified. Facilities for through overnight transits to the South of France, for example, should appeal strongly to many holiday-makers who find the long autoroute journey tiring and expensive in petrol, tolls, and hotel bills.

Freight traffic should, hopefully, be of trainload character as far as possible; but persuading the Continental administrations to adopt whole-heartedly the British Freightliner concept of block trains between a limited number of road-rail transfer terminals has been an uphill task. The Intercontainer consortium has made some progress, but individual containers as well as wagons are still too often (by British standards) being passed through marshalling yards, always a source of potential delay. There will almost certainly be a substantial SNCF marshalling yard at Fréthun. Still, a considerable number of through freight trains from British centres to major Continental destinations are to be expected. Trade cars in trainloads from manufacturers to dealers are another prospective traffic – sadly, more on the import than on the export side.

The routes of these planned services are conventional for the passenger trains; the present day No 1 boat train route via Sevenoaks and Tonbridge will be chiefly used, with the secondary route via Swanley and Maidstone East – some 12 minutes slower – as a relief facility. But freight trains from the North and West via the West London line will follow the main Brighton line as far as Redhill, joining the passenger route at Tonbridge. This route will be cleared structurally to handle 8ft 6in high ISO containers and all Freightliner trains will use it. Some other freight trains may however run via Swanley and Maidstone East, and thus avoid congesting the main line west of Ashford. There is yet another possible route via Swanley, Otford, and Sevenoaks, avoiding the busy No 1 route between Petts Wood and Sevenoaks, which may be helpful when, for instance, engineering work is required at night on the main line, especially in Polhill Tunnel.

All freight will debouch into the yard at Dolland's Moor before being cleared to enter the tunnel. The advantages of these routes are, largely, for keeping the busy two-track sections between Tonbridge and the West London line junctions near Stewarts Lane clear of freight as much as possible.

It will be asked how it is that the Southern Region, which in 1971 was so emphatic that pathways could not be found for Channel Tunnel trains, can apparently now provide them on a fully adequate basis. One answer is that over the Region as a whole, inner suburban services have declined following a fall in traffic. Some 400 carriages have in fact been taken out of the operating stock. Another is the improved track capacity provided by such resignalling schemes as that for the Victoria area (and the one in the early stages for Waterloo). These are bonuses over and above the obvious possibility of running tunnel trains in both the previous public timetable boat train timings, and also in all the optional, unadvertised alternative times, which are long-standing pathways in the working timetable. Lastly, perhaps one may suggest the appearance of enthusiasm for the tunnel's potential among the Region's management, after a long period of difficulty and struggle with domestic problems.

18 Light at the End?

For many years journalists have headed articles with either 'No Light at the End of the Chunnel' or, less frequently, 'Light at the End of Chunnel?' The apparently unending sequence of alternating hostility and support in Britain for this project will surely seem perplexing to future historians. (It has long been mystifying to the French.) Perhaps some British scepticism about the tunnel has been due to the fact that it was originally projected in an atmosphere of fantasy before the technology was ready to turn the vision into reality; and some of the derision aroused in early days, as well as some emotional, insular, habits of thought have persisted long after the technology became available to tackle the task. In fact, building the tunnel is today a problem much more of logistics and public relations than of technology. From about the 1920s onward it has been a soundly-based, robust concept from both the engineering and economic aspects.

It was perhaps a pity that Sir Edward Watkin became the tunnel's chief advocate in the 1880s. We can smile at the objections of the military, mentally preparing to fight the Napoleonic wars all over again. But Watkin's engineers had not really solved the problem of ventilation with steam traction. Colonel Beaumont's suggestion of compressed-air locomotives did not get very far; 1882 was not so long since the Metropolitan Railway had tried and failed to run trains (for a mere six miles) with a 'smokeless' locomotive – 'Fowler's Ghost', with steam-raising from hot bricks. Watkin is often remembered for grandiose planning – his Wembley Tower was intended to be taller than the Eiffel Tower, over 1,000ft high; but it never exceeded 200ft and, a financial failure, was eventually demolished with explosives. Watkin's Channel Tunnel workings did not suffer such a dramatic fate; but the stigma of failure hung around them.

And the modern tunnel project, endlessly researched and soundly based, undoubtedly suffered in the public mind in the 1970s by being linked, quite misleadingly, with two other projects, of a very costly character – Concorde and Maplin Airport. The association was utterly wrong because the tunnel was to be privately financed, unlike the other projects, and its economic prospects had been far more thoroughly studied. In the event, Maplin and the tunnel were the losers, in favour of Concorde – a good illustration of how dangerous it is to leave economic planning to politicians.

Once the tunnel is built, it will seem inconceivable that we waited so long for it. Of course it is right that environmental considerations should weigh heavily, at least as heavily as the financial interests of those who fear competition. But, on balance, the siphoning off of the ever-increasing volume of cars, coaches and heavy goods vehicles now pouring into Dover and Folkestone daily to reach the docks must be a major environmental gain. So too must be the diversion to rail and away from the motorways, especially the overtaxed M25, of road-borne containers and other freight. The scars caused by the construction sites will heal quickly, just as the cuttings and embankments of the railways, which caused protests when they were proposed, have sunk into the landscape. A researcher into the history of the tunnel in, say, the year 2050, will compare some of the opposition arguments with those of the early opponents of the railways, which nowadays we smile at, when we read their extravagant language and groundless fears.

To the names of the railway pioneers who refused to be intimidated by opposition, one would like to add those to Thomé de Gamond, Sir Arthur Fell, Sir William Bull, Leo d'Erlanger and countless others who have contributed to the struggle for the tunnel. This book ends with the same quotation as that at the beginning. 'Cessons d'écrire . . . creusons le sol'.

Appendix A

*including on the British side non-shareholding merchant bank advisers, Morgan Grenfell & Co, and Robert Fleming & Co

UK   *France*
                                EUROBRIDGE
                                   Bankers
Arbuthnot Latham   —
                                Contractors
Imperial Chemical Industries   —
Dupont Fibres   —
Brown & Root   —
John Laing   —
                        CHANNEL EXPRESSWAY
Sea Containers Ltd

                                   Bankers
—   Credit du Nord
                                Contractors
—   SCREC

Appendix B

*Some problems of the British and
Continental loading gauges*

The following notes are based, with his permission, upon a longer comprehensive paper by Mr C. A. Tysall, BSc, CEng.

Track gauge, the distance between the rails, is a relatively simple concept and there is no significant difference between the British and Continental track gauges other than a millimetre or so. Loading gauge however, (which determines vehicle dimensions) and structure gauge (which determines the minimum dimensions of bridge and tunnel openings, and how near to the track platforms and lineside equipment may be placed) are more complicated.

Where the track is curved, vehicles have an overthrow both at the centre and at the ends, and this increases the width of swept paths of vehicles on such curves.

British practice defines both the vehicle cross-section and the corresponding structure gauge on straight track, to provide adequate clearances. Where a curve is involved, the structure gauge is increased by the maximum overthrow from long vehicles such as passenger coaches.

The 'Berne gauge' which was adopted for European railways at a conference in 1913 as the standard for international services, prescribes the maximum swept width cross-section of rolling stock on a curve of 250 metres radius. This means that short vehicles such as four-wheel freight wagons can be built to the maximum width allowed, but passenger coaches must be correspondingly narrower because of their greater length.

Passenger rolling stock in Britain has been designed to the C1 gauge for many years. This prescribes the maximum body width of coaches as 9ft 0in (9ft 3in over door handles) or 9ft 3in body width, in the case of sliding-door stock without further projections. BR's modern Inter-City Mk III coach is designed to run where C1 stock is permitted, but its greater length of 75ft 0in (C1 stock is normally 63ft 6in long over headstocks or less) requires additional lateral clearances where the radius is very small.

The UIC (International Union of Railways) has established a range of standard vehicle dimensions conforming to Berne gauge limits. While short freight wagons are nearly 10ft 4in wide, two standard passenger coach types are narrower, as the following table shows:—

UIC type	Length	Width
X	80ft 5in	9ft 6in
Y	86ft 9in	9ft 3in

While the height of BR stock must not exceed 12ft 10in, the height of Berne gauge stock can be a maximum of 14ft 0½in. It appears that certain types of Continental stock such as the UIC X type could run on a number of Southern Region tracks with only relatively minor adjustments to lateral clearances; but finding an additional 14½ inches of vertical clearance would involve some lifting of overbridges or lowering of track, or a combination of both, at each restricted site. The provision of special Channel Tunnel stock is therefore probably the more economic solution; it would not only conform to BR's tighter clearances but can be built to accommodate the various platform heights on the European administrations served.

Index